I0055640

A DIALOGUE CONCERNING THE TWO CHIEF MODELS OF PLANET FORMATION

A DIALOGUE CONCERNING THE TWO CHIEF MODELS OF PLANET FORMATION

Michael Mark Woolfson

University of York, UK

World Scientific

NEW JERSEY · LONDON · SINGAPORE · BEIJING · SHANGHAI · HONG KONG · TAIPEI · CHENNAI · TOKYO

Published by

World Scientific Publishing Europe Ltd.

57 Shelton Street, Covent Garden, London WC2H 9HE

Head office: 5 Toh Tuck Link, Singapore 596224

USA office: 27 Warren Street, Suite 401-402, Hackensack, NJ 07601

Library of Congress Cataloging-in-Publication Data
Names: Woolfson, Michael M. (Michael Mark), author.
Title: A dialogue concerning the two chief models of planet formation /
    by Michael Mark Woolfson (University of York, UK).
Description: Singapore ; Hackensack, NJ : World Scientific, [2017]
Identifiers: LCCN 2017001714| ISBN 9781786342720 (hc ; alk. paper) |
    ISBN 1786342723 (hc ; alk. paper) | ISBN 9781786342737 (pbk ; alk. paper) |
    ISBN 1786342731 (pbk ; alk. paper)
Subjects: LCSH: Solar system--Origin. | Planets--Origin. | Nebular hypothesis.
Classification: LCC QB503 .W657 2017 | DDC 523.4--dc23
LC record available at https://lccn.loc.gov/2017001714

British Library Cataloguing-in-Publication Data
A catalogue record for this book is available from the British Library.

Copyright © 2017 by World Scientific Publishing Europe Ltd.

All rights reserved. This book, or parts thereof, may not be reproduced in any form or by any means, electronic or mechanical, including photocopying, recording or any information storage and retrieval system now known or to be invented, without written permission from the Publisher.

For photocopying of material in this volume, please pay a copying fee through the Copyright Clearance Center, Inc., 222 Rosewood Drive, Danvers, MA 01923, USA. In this case permission to photocopy is not required from the publisher.

Desk Editors: Dipasri Sardar/Mary Simpson/Shi Ying Koe

Typeset by Stallion Press
Email: enquiries@stallionpress.com

Introduction

The history of science abounds with disputes concerning the relative merits of alternative theories and astronomy is a particularly rich source of such controversy. An early astronomical example was about the motions of bodies in the Solar System. From the 2nd century through to the 17th century, Ptolemy's theory — that all the bodies in the Universe revolved around a stationary Earth — was accepted by the vast majority of astronomers. The bodies assumed to be orbiting the Earth included the Sun, Moon and planets. This model was also strongly supported by the Church, both Catholic and Protestant, because it was in accord with the Biblical account of the creation and the special nature of the Earth and mankind. In the mid-16th century an alternative model, that the bodies of the Solar System all orbited the Sun, was proposed by the Polish Catholic cleric, Nicolaus Copernicus. Despite the fact that it better agreed with observations and gave a much simpler and more convincing model of the Solar System, it took more than a century for it to be universally accepted. Once an idea becomes entrenched and has wide acceptance it takes a long time for an alternative to be considered, let alone accepted.

The next example, not from the field of astronomy, concerns the nature of light. In the 17th century there was a dispute, involving very eminent scientists, about whether light consisted of a stream of particles or whether it was some kind of wave motion. Isaac Newton considered that light consisted of corpuscles — small particles — but the Dutch scientist Christiaan Huygens favoured the idea that it was a wave motion. The dispute seemed to be settled in the 18th century when Thomas Young showed that light could be used to produce an interference pattern, something that could only be done with waves. However, the story was not finished. In the early 20th century Albert Einstein theoretically explained the results of experiments on the photoelectric effect — the emission of electrons from a metal when light shines on it — that showed clearly that light was behaving like particles — given the name *photons*. Considering the fierce debate about the nature of light it is rather curious that we now know that light can manifest the properties of *both* particles *and* waves, depending on the experiment or application in which it is involved. This idea of wave-particle duality also applies to electrons that are normally thought of as particles with a particular mass and charge but can reveal their wave nature when producing an image using an electron microscope.

A final historical example, from the 20th century, is the rather acrimonious dispute between Martin Ryle and Fred Hoyle concerning the evolution of the Universe. Ryle, a radio astronomer, believed that, as the Universe expanded, so the density of matter within it would fall, which seems intuitively obvious. Hoyle, a theoretical astronomer, supported the 'Steady State Universe' in which there was a 'creation field' that created matter so as to keep the density of matter in the Universe constant. By measurements of the density of radio sources at different distances Ryle's model was eventually verified.

A modern controversy concerns the way in which planets are, and, in particular, the Solar System was, formed. However, the fact that there *is* a controversy is not widely known; one side of the argument has a great following and the other side a very small one — the ex-members of my research group and the occasional reader of my work who contacts me to say that they found it persuasive. Now, having retired, I work alone and a lone voice in the midst of a noisy hubbub is not easily heard especially as, through infirmity, I am unable to attend conferences and engage in face-to-face discussion. I can only communicate my ideas in print and such material must surmount the barrier of a sometimes antagonistic review process — but I will not dwell on that.

Ideally, in presenting theories for comparison each should be presented by a proponent who would present the strongest possible case. Unfortunately, despite many attempts, I have never been able to find a partner who would present the case for the Nebula Theory, the so-called 'standard model' in conjunction with my presentation of the Capture Theory. So I have done my best to present both cases in the best possible light. I believe, and hope, that I have been objective and honest.

Prologue

When observing the night sky most of what is seen is a myriad of stars and, just within the limits of visual acuity, some fuzzy looking 'stars' that are now known to be complete galaxies. The positions of the stars are fixed relative to each other and they have been arbitrarily grouped together to form constellations to which names have been given such as *scorpio* (the scorpion) and *cancer* (the crab). As the night progresses so these constellations rotate around a point in the sky, marked by the Pole Star in the northern hemisphere. Since *we* are not moving — otherwise, surely we would sense the motion in some way — it is clear that we on Earth are at the centre of the Universe, which spins around the Earth once per day. So it seemed to the ancients and this Earth-centred model, due to the Alexandrian Greek philosopher Ptolemy (100–170 AD), was unchallenged for 1,400 years and was accepted by most astronomers for even longer than that without question. Early Greek philosophy held the view that nature espoused perfection and, since a circle was regarded as a perfect shape, it was thought that all heavenly motions should be circular around the Earth at a constant speed. Hence, it was found troubling that some planets occasionally reversed their apparent motion around the Earth but this could be explained by introducing the idea that a planet made two coupled circular motions, i.e. an *epicycle*, a circular motion at a constant angular speed around a point, the *deferent*, that itself orbited the Earth in a circular orbit at a constant angular speed (Figure P1). The motion was complicated but, since the laws that governed the motions of heavenly bodies were not known at that time, so be it. This, the *geocentric model*, was the standard model in the early 16th century. The Earth was stationary and all other bodies, including the Sun, moved around it at a constant speed, with the proviso that for planets the motion consisted of two coupled circular motions at a constant speed. The model had the approval of both the Catholic and Protestant Churches. In the Biblical account of creation, God had first created 'the heavens and the earth' and finally 'mankind in his own image', so that having the Earth, inhabited by mankind, at the centre of all creation seemed natural and proper.

Nicolaus Copernicus (1473–1543; Figure P2) was a Polish Augustinian canon, born in the city of Torun, and a man of many talents — in mathematics, astronomy, medicine, economics and politics. He made improved astronomical observations of planetary motions from which he concluded that a description of the Solar System with planets, including the Earth, all orbiting the Sun — a *heliocentric model* — was more consistent with the observations. In 1543, he described his model in a book, *Revolutionibus Orbium Coelestium* (*On the Orbits of Heavenly Bodies*); it is said that he received the printed version of the book on his death-bed. Copernicus was still reluctant to abandon the idea that all motions had to be circular at a constant speed so, since planetary orbits are actually ellipses, he assumed that the Sun was slightly displaced from the orbital centres, which gave varying distances from the Sun, as observed, and also explained why planets have a greater angular speed when closer to the Sun. Even so he still had to introduce small epicycles to match his observations. This heliocentric system was much simpler than Ptolemy's geocentric one but it failed to gain much acceptance in the astronomical community; a belief held for 1,400 years by many generations of scientists was not to be easily abandoned.

At the time of publication of *De Revolutionibus* the Catholic Church, the most powerful institution of its day in most of Europe, seemed to be indifferent to its message; in fact the book was dedicated to Pope Paul III. In any case, since it was mostly disregarded

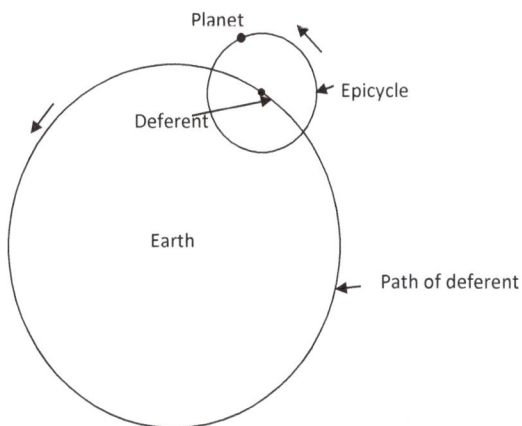

Figure P1   Ptolemy's description of planetary motion

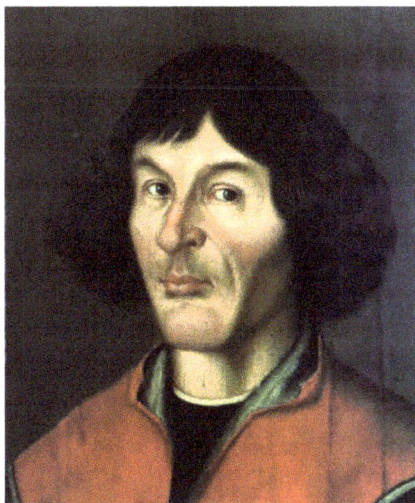

Figure P2   Nicolaus Copernicus

it did not really challenge the Church's fundamental belief in the geocentric model; they could afford to be tolerant of individual eccentricity. However, more than half a century after Copernicus died this indifference was destined to change, triggered by the activities of a Dominican monk and philosopher, Giordano Bruno (1548–1600; Figure P3). He had a number of heretical views including his denial

Figure P3   Giordano Bruno

of the divinity of Christ, the virginity of Mary and the idea of Eternal Damnation — one cannot help feeling that he was in the wrong profession. He believed in an infinite Universe so that no location could be regarded as its centre and he also proposed that stars were just like the Sun with orbiting planets inhabited by other races of men. This last proposition challenged one of the basic tenets of the Church concerning the central role of God's supreme creation, mankind, and for the Inquisition this was the final straw. In 1593, Bruno was transferred to face the Inquisition in Rome, having successfully defended himself when appearing before the Venetian Inquisition. After seven years of an extended trial, during which he was imprisoned, he was found guilty of heresy and, since he would not recant, he was burnt at the stake in 1600. One of the principle charges brought against him was his belief in the plurality of worlds, linked to the Copernican heliocentric model of the Solar System.

Indisputably, the greatest scientist of his period was the Italian, Galileo Galilei (1562–1642; Figure P4) who made seminal contributions to both the theory of motion — kinematics and

Figure P4   Galileo Galilei

dynamics — and to astronomy. Indeed, he is very often given the accolade of being 'the father of observational astronomy'. Early in the 17th century the telescope was invented, probably by the Dutch spectacle maker Hans Lippershey (1570–1619). In 1608 Galileo constructed a telescope that gave a tenfold magnification, much greater than the original Dutch version could attain, In developing his telescope he had the encouragement and support of the Venetian Senate, to which body the ability of being able to see ships approaching the harbour while still two hours sailing-time away had commercial and military value. Galileo used his telescope to examine various heavenly bodies. He saw mountains on the Moon and estimated their heights by the lengths of the shadows they cast. He found the large satellites of Jupiter — Io, Europa, Ganymede and Callisto, now known as the *Galilean Satellites* in his honour, and he saw the rings of Saturn, although the resolution of his telescope was too poor for him to interpret what he saw.

The most significant observations Galileo made were those of the phases of Venus, observations that convinced him that the

Copernican model had to be correct. The observation was that Venus
could sometimes be seen as a large crescent and at other times as a
small disk. Venus is always seen in the same general direction as the
Sun and Ptolemy's model explained this by having the deferent of
Venus always aligned with that of the Sun. As Venus went round
its epicycle it was first seen on one side of the Sun and then on
the other. However, for this geocentric model it was always between
the Earth and the Sun so it was the side of Venus furthest from
Earth that was always illuminated and Venus should then only
be seen in crescent phase from light coming from an edge of the
illuminated hemisphere (Figure P5(a)). By contrast, on the basis of

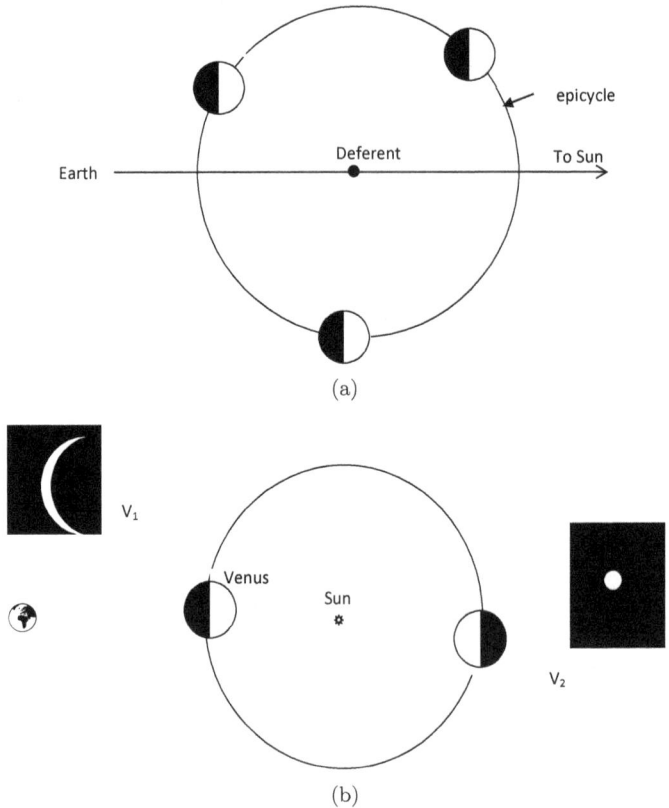

Figure P5   (a) Ptolemy's arrangement of the Earth, Venus and the Sun. (b) The
heliocentric arrangement

the Copernican heliocentric model the orbit of Venus was within that of the Earth so that Venus could be close to the Earth and between the Earth and the Sun (position V_1, Figure P5(b)), and so seen as a large crescent, or on the far side of the Sun from the Earth so that the whole illuminated hemisphere could be seen as a disk, but much smaller in size than the crescent (position V_2, Figure P5(b)).

In 1610, Galileo published his observations of Jupiter and Venus in *Sidereus Nuncius* (The Starry Messenger), which promoted the heliocentric model of the Solar System and this once again provoked opposition by the Church. In 1616, support for the heliocentric theory was declared as heresy and all heretical publications, which included works by Copernicus, Johannes Kepler and Galileo, were placed on the *Index Librorum Prohibitorum,* the list of books that Catholics were forbidden to read. Thereafter, to openly support the Copernican model was very hazardous indeed. Copernicus never knew of the furore that his ideas were to produce.

This put Galileo in a quandary since he had no wish to run afoul of the Church; apart from the danger of doing so he was a pious man, despite having fathered three illegitimate children, but the scientific evidence was irrefutable — the geocentric theory was wrong and the heliocentric theory correct. What could he do to proclaim the truth as revealed by science against the all-powerful opposition of the Church? He decided on the device of writing a book, *Dialogue Concerning the Two Chief World Systems* (Figure P6), in the form of a dispassionate discussion of the relative merits of the geocentric and heliocentric models by two individuals — Simplicius, supporting the geocentric model, and Salviati supporting the heliocentric model. Also present was a neutral observer, Sagredo, who listened to the cases being made and acted as a catalyst for the discussion by asking questions from time-to-time.

Galileo's plan failed abysmally. It was clear to any intelligent reader which of the two individuals — Simplicius and Salviati — had made the better case and the members of the Inquisition were nothing if not intelligent. Galileo had to appear before a tribunal and asked *to abandon completely the opinion that the sun stands still at*

Figure P6   Frontispiece and title page of the *Dialogue*, 1632

the centre of the world and the earth moves, and henceforth not to hold, teach, or defend it in any way whatever, either orally or in writing.', something that, as a sensible man, he said he would do — he was no Bruno! It is rumoured that he muttered 'eppur si muove' — 'but still it moves' — under his breath immediately after recanting his heretical views. Despite his submission, Galileo was sentenced to house arrest for the rest of his life but he escaped without physical harm.

The brilliant German astronomer and mathematician Johannes Kepler (1571–1630; Figure P7(a)), using the accurate planetary observations of the Danish astronomer Tycho Brahe (1546–1601), showed that planetary orbits around the Sun are ellipses and, finally, when Isaac Newton (1642–1727; Figure P7(b)), probably the greatest scientist of all time, showed that the inverse-square law of gravitation completely explained the nature of planetary orbits, the question of which model was correct was put beyond dispute.

This was a sorry chapter in the history of science when brute force was used to suppress the conclusions drawn from scientific

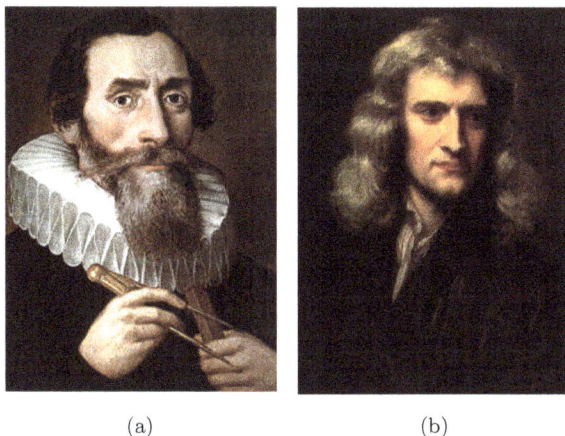

(a) (b)

Figure P7   The giants of solar-system theory (a) Johannes Kepler (b) Isaac Newton

observations. Unfortunately, it was not the last time a scientific theory was opposed by religious institutions on the basis that it was in conflict with the implications from biblical accounts. The more recent case was the opposition to Darwin's theory of evolution as described in *On the Origin of Species.* Fortunately such clashes between science and religion are no longer an issue, although there are still some who believe that the Earth is about 6,000 years old and others who believe that the Earth is flat.

In the 21st century there is another clash of ideas involving the Solar System, this time about how it was formed. Because of observations of planets around other stars, known as *exoplanets*, (well done Bruno!) these ideas are now concerned with general planet formation.

The first of these theories, originally known as the Solar Nebula Theory but now, because of the existence of exoplanets, the Nebula Theory (NT) has roots going back to 1796 when the French astronomer, Pierre Simon Laplace (1749–1827), in his book *Exposition du Système du Monde*, described the first scientifically-based theory of Solar System formation, based on the evolution of a nebula, a diffuse spinning sphere of dusty gas. This theory was generally accepted until the mid-19th century when it was abandoned because of a problem concerned with angular momentum. The Sun, with 99.86 percent of the mass of the system, possesses only 0.5 percent of its angular momentum. The remaining tiny fraction of the total mass has 99.5 percent of the angular momentum, contained in the planetary orbits. No way of partitioning mass and angular momentum in this way could be envisaged at that time.

In the 1960s there was a revival of the idea that the Sun and planets could have evolved from a nebula and now the NT has become established as the standard model for the formation of stars and their accompanying planets. The NT is accepted without question by the great majority of planetary scientists, although most of them are concerned with general aspects of the Solar System — planets, comets, meteorites, planetary atmospheres etc. — but not specifically with the origin of planetary systems. A vast amount of work is carried out in attempting to solve the many problems of the NT — problems that increase in number as new observations are made — and generating different solutions to the problems produces a large body of literature that adds to the impression of positive progress and hence to increased acceptance of the theory. Despite the problems, which are many and sometimes critical, belief in the NT is undiminished. It is the story of the *Emperor's New Clothes* all over again; those who do not see them are either stupid or incompetent, so everyone sees and admires the invisible clothes. So it is with belief in the NT. The model is the only one described in popular literature and in both radio and television popular programs and so has achieved the status of unchallengeable dogma for the intelligent reader, listener or viewer. Those who would believe otherwise must be mistaken. To break this spell someone must play the role of the child who proclaimed 'But he isn't wearing anything at all'.

Another theory put forward by me in the 1960s is the so-called Capture Theory (CT), which has tenuous links with a tidal theory put forward in 1917 by the British astronomer James Jeans (1877–1946). The NT is a *monistic theory* in which the star and planets derive from the same body of material. By contrast the Jeans theory and the CT are *dualistic theories* in which the star and planets are derived from different bodies of material. The Jeans theory became the widely-accepted standard model for two decades until some problems of the theory were identified — briefly, that planets could not form sufficiently far from the Sun to explain the present Solar System and, crucially, that planets could not form at all. The model became untenable but what was left behind was a body of theory, developed by Jeans, that has stood the test of time and is applied in many astronomical contexts. The CT draws heavily on this theory.

Unlike the NT, which generates problems, the CT generates solutions and scientifically credible explanations of observations, some of which have been declared to be *very challenging* by the observers themselves. However, that claim for the CT is just a statement and, without justification by scientific argument, it has no value. So, what follows is a modern version of Galileo's attempt to persuade the reader that one model is better than the other — in this case that the CT is more plausible than the NT. Note well the word 'plausible'; no theory can be claimed to be correct for, at any time, new analysis or a new observation may show that is it wrong. In this new *Dialogue* the presenter of the NT will be Simon, who will make the strongest case for the NT as it appears in the current literature. The CT will be presented by Steven who, again, will be mainly presenting what is in published material. The intelligent layman will be Solomon, the namesake of the biblical king who was deemed to be a wise judge. The discussion between Simon and Steven will include a full description of the theories of Laplace and Jeans, to which the NT and CT are related, and be carried out at a level that should be accessible to those with a reasonable school background in science. For those wanting more than that, references to literature, in which the science is explained more formally and in greater detail, will be given in footnotes.

This treatise is being written by one with a strong belief that the CT has far greater plausibility than the NT. However, the upmost attempt will be made to present the NT case as objectively and completely as possible — indeed, to act like a professional lawyer who presents the strongest possible case for his client, even if he believes the client to be guilty. In this sense, my role is unlike that of Galileo whose partisanship in writing his *Dialogue* was evident throughout. However, again unlike Galileo, I have no threatening Inquisition in the background so my physical welfare is not threatened although, if the case for the CT is deemed to be flimsy, my reputation as a scientist may be put on the line. Whether or not that case is made is up to the judgement of the reader.

A Chance Encounter

The scene is a street in central London on 22nd January 2016. Two young men, coming from opposite directions, meet and stop to talk.

Simon: Steve, how are you? It's more than 5 years since graduation day when we were last together. Do you remember the party we had that night!

Steven: I certainly do — just the memory of it brings on a hangover! I'm fine thanks. I hardly recognized you with the beard, but it suits you and you seem hale and hearty. What are you doing now? What's wrong, you look startled, what are you looking at?

Simon: You are not going to believe this. Do you remember Solomon — Sol we called him but I can't recall his surname — the bright theoretical chap in our year? He got the prize for being the best theoretician. He's coming our way from behind you. Sol, do you remember me?

Solomon: Simon, of course I do despite the foliage and Steve as well. Fancy meeting you two after all this time. You've obviously kept in touch.

Simon: It's even stranger than you think. Steve and I only met by chance about 3 minutes ago. You're the theoretical wizard Sol. What's the chance of three people who haven't met for 5 years all coming together by chance at the same place at the same time?

Solomon: That's not a problem I can solve — too little data — but tell me what are you both doing now?

Simon: I'm a Research Fellow in the astronomy department at St Jude's, working with Prof. Burton on planet migration. He's a big wheel in the nebula-theory community.

Steven: A big windbag you mean. That theory is a dead as the dodo but it just refuses to lie down. It's a scientific zombie.

Simon: What makes you say that? Anyway what are you doing now?

Steven: Well, as it happens I am also working in the planet-formation field, as an assistant to Dick Turnbull. We are looking at the formation of the Oort Cloud on the basis of the capture-theory model.

Simon: You can't be serious. That model is complete rubbish. Turnbull and his crew just ignore the fact that everyone but them knows that the Nebula Theory is the only viable model at the present time. They act like the soldier in the marching column who says that everyone else is out of step.

Solomon: Hey you two — I don't want to intrude on a private battle but I am intrigued. Why do planets migrate and what is the Oort Cloud? I just know some general stuff about planets and the Solar System but what you are talking about makes no sense to me. As far as I know planets stay in their orbits and clouds belong in the sky.

Simon: The nebula-theory model produces solar-system planets closer to the Sun than they are now and through interactions with a gaseous disk containing asteroid-size bodies and with each other they migrate outwards to their present positions.

Steven: The Oort Cloud consists of a vast number of comets that spend most of their time far from the Sun — tens of thousands of times the Earth's distance from the Sun. Occasionally one of them is perturbed into the inner Solar

System and from the observed positions of the comet over time we can determine whence it came. But Simon, what do you actually know about the Capture Theory? Very little I suspect.

Simon: Well, I see occasional papers in the main astronomical journals and I read the abstracts but there are many way-out ideas about planet formation that somehow get published, including capture-theory ideas, and I really can't waste my time reading this stuff that has no attachment to reality. The Nebula Theory is so well established and so soundly based I cannot believe that anyone would want to spend time working on anything else.

Steven: So, what you are saying is that you really know next to nothing about the Capture Theory — yet you are making statements about its scientific credentials. Is that a good scientific attitude? For my part I think that the Nebula Theory is about as watertight as a sieve — full of holes that can't be plugged. That statement isn't made from ignorance about the Nebula Theory but from knowledge about it. Unlike you, I read everything relative to planet formation and keep an open mind.

Solomon: You two are bickering like a pair of politicians or fanatical theologians with different views. I work in particle physics and we have our spats from time-to-time but we discuss them scientifically. You, Simon, are saying that since the vast majority of people working in the field believe in the Nebula Theory then it must be true. On that basis, if you had lived in the 16th century you would have been a strong adherent to Ptolemy's geocentric model of the Solar System. On the other hand you, Steven, lightly dismiss the fact that a large number of bright and distinguished people work on the Nebula Theory — as Simon remarked, you are saying that everyone else is out of step.

Steven: Sol, you have no idea how frustrating it is. You write a paper based on sound scientific principles, you have a battle getting through referees who, because they numerically dominate the field, are almost certainly going to be in favour of the Nebula Theory, and having done all that you produce a paper that is largely ignored. That seems to be the antithesis of what science should be about.

Simon: You're exaggerating, Steven. Plenty of capture-theory stuff gets published; it is just that it isn't found convincing.

Steven: By your own admission you haven't read any capture-theory papers, so how could you be convinced? Look, let's get this straight. By the nature of this field no theory can ever claim to be correct, and I include the Capture Theory in that statement. The only measure is plausibility and that is time-dependent since there are new observations made from time-to-time that test existing theories in different ways. Some of the most recent observations have been accompanied by observer comments that they are challenging to theories of planet formation, by which they mean theories based on nebulae since they know of nothing else. Well, I can tell you that these recent observations are *not* challenging to the Capture Theory.

Solomon: Look you two, this sort of argy-bargy can go on forever. I have just had a thought. I get the impression that what you are both doing is interesting and I would like to know more about it. I read a book a couple of years ago about scientific disputes including those involving the geocentric or heliocentric Solar System, the nature of light — waves or particles — and the steady-state versus non-steady-state Universe. However, you have just introduced me to a new one that I have never heard of.

Let us put this serendipitous meeting to some good purpose. Why not meet once a week, say in a pub, and you can explain to me, and to each other, aspects of what you are doing and the basis of the two theories. Not only would I find it of interest but, by the process of explaining it to me, a non-specialist, you might clarify what you are doing in your own minds. You could bring with you figures, photographs and tables to illustrate what you are saying.

Simon: Seems a good idea and I'm up for it. It might be fun and it would be good to renew our acquaintanceship. Although Steve and I both feel strongly about the validity of the work we are doing this has nothing to do with personal relationships and I am really pleased to have met you both.

Steven: I echo everything Simon has just said.

Solomon: Well then, what about next Friday at, say, 8 o'clock. Can you both manage that?

Steven: OK for me.

Simon: And for me.

Solomon: Do you know The Anchor in Belham Street? It's my usual watering-hole and they serve decent real ale. In addition, they have cubicles so if we can get one of those we can talk without being interrupted and without disturbing others.

Simon: I know it — it has a good atmosphere.

Steven: I don't know it but I'll find it.

Solomon: Good. I suggest that next week you just concentrate on explaining the basic planet-formation process according to the two theories. That will be a good starting point and, I am sure, will lead into many other areas of discussion. I must go now. My fiancée, Alice, is always on time and she will be most unpleased if I am late for our date. Bye you two.

Steven: Bye Sol. See you next week. Simon, I too must dash to catch a train to Oxford. I look forward to our discussions.

Simon: Me too. Have a good trip.

The three leave on their separate ways.

Contents

The First Meeting

Simon Describes the Basic Mechanism
of the Nebula Theory

On 29th January 2016, the three friends are in a cubicle in The Anchor public house. Simon and Steven are sitting side-by side and Solomon is on the opposite bench, strategically placed in the centre so as not to be immediately opposite either of the other two. Three pints of real ale, provided by Solomon, are on the table, each already having been sampled with much pleasure by the drinkers.

Steven: Did you get to your date on time last week, Sol?

Solomon: I was a minute late but Alice was in a forgiving mood so the engagement is still on. Anyway, down to business. I've given some thought as to how to get this process going so what I suggest is that we start by each of you describing the basic mechanism by which your favoured theory produces planets. Let's not worry about any problems there might be for now; I'll possibly spot some but I am sure that each of you will be only too willing to point out difficulties with the other theory. Is that OK with you? Take your time and explain things clearly. While I know a great deal about particle physics I have only a general knowledge of your field so keep it simple and don't leave any gaps.

Simon: Yes, that's a good idea. We'll keep it simple at first and get to the difficulties later on. There's just one thing though. I imagine that we don't want to be here all night and when I assembled my presentation material I realized that to put forward a comprehensive and comprehensible account would probably take all the time we would wish to stay here this evening. What about you, Steve?

Steven: I thought the same. While presenting the basic capture-theory mechanism of producing planets may take less time than presenting your stuff I don't think it is practicable to present them both in one evening. Anyway, since the Nebula Theory is the 'standard model' I think that you should kick off. I promise not to interrupt so you can have an uninterrupted run; Sol won't be able to get a clear picture if we keep interrupting each other and the narrative gets side-tracked away from the main message.

Solomon: So, we're agreed — just Simon this week and Steve next time. Off you go Simon. Did you bring some material to illustrate your description of planet formation? You know the old adage — a picture is worth a thousand words.

Simon: Yes, I have a few bits and pieces. Well here goes. To start my story I have to go back over 200 years. In 1796 the French astronomer, Pierre Simon Laplace, seen in this portrait (Figure 1.1), gave what was probably the first scientifically-based theory of how planets might form.[1] You can follow the process in this figure (Figure 1.2). Laplace postulated that the Solar System began as a slowly-spinning cooling sphere of dusty gas that was gradually collapsing under the influence of its own self-gravity. To conserve the angular momentum of this isolated body, as it collapsed it had to spin faster and it flattened along the spin axis, as seen here in frame (a). Eventually, the angular speed of material at the equator was high enough for it to be in free orbit around the central mass. At that point, the body of gas had taken on a lenticular shape; you can see this in frame (b); it is like a biconvex optical lens but with a sharp edge. The centre continued to collapse

---

[1]Laplace, P.S. de (1796) *Exposition du Systéme du Monde* (Imprimerie Cercle Social, Paris).

Figure 1.1   Pierre Simon Laplace (1745–1827)

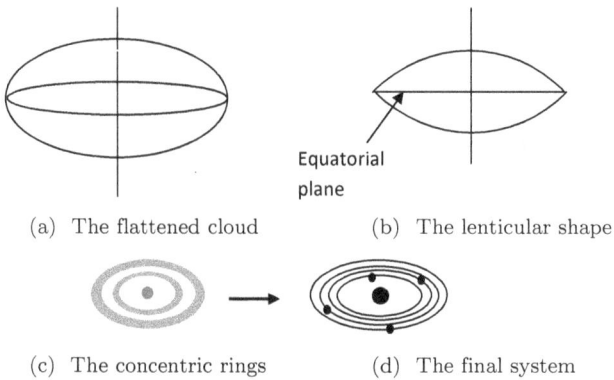

(a) The flattened cloud

Equatorial
plane

(b) The lenticular shape

(c) The concentric rings

(d) The final system

Figure 1.2   The Laplace nebular model

and as it did so material further in from the boundary
progressively acquired free-orbit angular speed, the effect
being that a disk of material formed in the equatorial
plane. Laplace assumed, without giving any explanation,
that disk material detached from the collapsing core in
a spasmodic fashion so that the layer of material left

behind did not form a continuous disk but rather a series of concentric rings, as you see in frame (c). Eventually, the central core of material collapsed to form the Sun and material began to coalesce in the rings so that each ring contained a number of blobs of material. Within each ring, individual blobs were in slightly different orbits and had slightly different orbital speeds so that faster blobs caught up with slower ones and they combined. Eventually, in each ring a single blob of material was left and, in due course, this collapsed to form a planet, as seen in final frame (d).

This theory explained what was known about the Solar System at that time. It gave all planets orbiting in the same sense and approximately in the same plane. However, by the mid-19th century the viability of the theory was being questioned because of problems concerned with the distribution of angular momentum in the Solar System. The Sun has 700 times as much mass as all the planets combined and the planets in their orbits have 200 times as much angular momentum as the Sun has in its spin; the Sun spins very slowly with an orbital period of about 28 days. There were other ways of expressing this problem. For example, if the original sphere had the angular momentum of the whole Solar System and was spinning uniformly like a solid body then, when it had reached the radius of Neptune in its collapse, material at its equator would have much less than the angular speed to be in free orbit so a Neptune ring could not form. The same would be true for some of the other outer planets. Again, expressing the problem in another way, if when Mercury formed, the inner core had the orbital angular speed of Mercury then it would have about 2,000 times the present angular momentum of the Sun — so what happened to all that angular momentum?

There were attempts to rescue the theory by postulating extreme distributions of material in the original nebula. A French astronomer, Eduard Roche (1820–1883), considered an original nebula in which most of the material was concentrated at the centre with very little in the outer regions. This certainly improved the situation with respect to the mass-angular momentum distribution — although it was still pretty bad — but it meant that there was too little material in the outer part of the nebula to produce planets at all! The various arguments against Laplace's nebula model were so strong that by the beginning of the 20th century all attempts to rescue it had virtually ceased and the scientific community was receptive to new ideas.

Between the middle of the 19th century and the middle of the 20th century many ideas were suggested for the formation of the Solar System; the only one that had acceptance for a while was the tidal model proposed by James Jeans in 1917 but even that was eventually found to be wanting. I suppose that Steve will tell you about that one because it has some links with the Capture Theory.

Steven: Yes, Simon, I will be describing Jeans' model.

Simon: Now we move forward to the 1960s. In meteorite research there are many indications that the material of some meteorites had, at some time, existed as a vapour. When silicate minerals are heated to such a high temperature that they vaporize, the chemical structures of the minerals break down into units that are stable at the prevailing temperature. For example, the mineral *forsterite*, Mg_2SiO_4 will break up into $2MgO + SiO_2$. If now the original rock was a mixture of many minerals then the vapour will contain a mixture of the stable groups derived from the different minerals. When this vapour subsequently cools, the first minerals that condense out of the mixture, at first in liquid form, will be those with the highest vaporization temperatures, which could even consist of minerals not present in the original rock. The inner parts of the grains of some meteorites consist of minerals that condensed at high temperatures and, as one moves outwards from the centre, the condensation temperatures of the minerals decreases.

This suggests successive condensation in a cooling vapour. Another indication of previous vaporization is that in some meteorites there are cavities with crystals on the cavity walls that seem to have condensed directly from the vapour phase to a solid, the reverse process of sublimation where a solid converts directly into a vapour. The presence of a hot vapour in the early Solar System led to the idea that it could have been a hot cooling nebula in association with a newly-forming star and work began to consider how such a system could give rise to the Sun and planets. It was felt that new knowledge would enable the problems of the Laplace model to be solved. It was soon realized that if the nebula had been hot enough to vaporize silicates then you would not be able to produce planets from it directly — it is more-or-less the argument raised against Jeans' tidal model by the American astrophysicist Lyman Spitzer but I am sure that Steve will tell you about that later. Basically, planets could only form when you have a central star with a surrounding cool disk — very much the same starting point that Laplace used. That is no barrier to a nebula theory; hot material would eventually cool and then planets could form.

You would have spotted that I am apparently ignoring the angular momentum problem that bedevilled Laplace's model but I'll bypass that for now. Steve, or you Sol, will bring this up as a difficulty with the nebular model and I'll deal with it then. Most of the early work concentrated on the problem of how planets will form in a disk assuming that there was a disk but, as happens sometimes in science, assumptions made in developing a theory are later justified by observations, and this was so in this case.

Steven: Just a little interruption to confirm what Simon is saying. I'll be giving an example where later observations have justified an earlier assumption.

Simon: Thanks Steve. So I'll reverse the timing by explaining what the evidence is for the existence of disks before

Figure 1.3   The Orion Nebula

describing the processes that happened in the disks to produce planets. To be systematic about this I'll need to say something about the way that stars form. You did stress, Sol, that you wanted a complete picture.

When we look within the galaxy we find regions, rich in dust, where the average density of gas is about 1,000 times higher than in the galaxy as a whole and the temperature is very low, just a few tens of degrees kelvin (K). These are star-forming regions; one such is the Orion Nebula seen here (Figure 1.3) and within it various clusters of stars are forming.

The first detectable state of a new star is when it is in the form of a protostar, a large diffuse cool body of stellar mass and with radius of order 1,000–2,000 astronomical units, shortened to au. One au, which is 1.496×10^8 km, is the average distance between the Earth and the Sun. Protostars are very cool, with temperatures about 10–20 K but although bodies at that temperature are poor radiators of energy in terms of output per unit area, and radiate in the far infrared part of the spectrum, because they have such a large surface area the total radiation from them can be detected by infrared detectors.

A very useful way of envisaging the state of a star is by plotting it on a Hertzsprung–Russell (shortened to H–R) diagram, devised independently by the Danish astronomer Ejnar Hertzsprung (1873–1967) and the American astronomer Henry Norris Russell (1877–1957). In one version of this diagram the temperature of the star is plotted on the x-axis and the luminosity, i.e. the total energy output per unit time, on the y-axis. Here is an example of a H–R diagram (Figure 1.4). You can learn a lot about a star from its position on the diagram. For example, if you have a plotted star at a fairly low temperature, say 3,000 K, but it is 1,000 times as luminous as the Sun, then from the temperature, you know that it mainly radiates in the red part of the spectrum and that it is a fairly poor radiator of energy per unit area. Then, from its high luminosity, you know that it must be very large — the kind of star known as a *red giant*. By the same kind of reasoning a star at a high temperature, say 12,000 K, but with a luminosity one-hundredth of that of the Sun must be a very small star — one known as a *white dwarf*. You will see from this diagram that most observed stars fall on the curved line marked as *main sequence*. These are stars that have reached the stage where they are converting hydrogen to helium in their cores and for stars of moderate or low mass this is a long-lasting state. The Sun has been on the main sequence for about 5 billion years and will remain on it for another 5 billion years. All main sequence stars with the same temperature — and hence luminosity — are similar so from the position on the main sequence everything is known about the star — not only the plotted quantities but also its mass and radius.

A new protostar, with its very low temperature, occupies a place well off the diagram to the right of the main-sequence line. In the initial stage of its collapse it stays

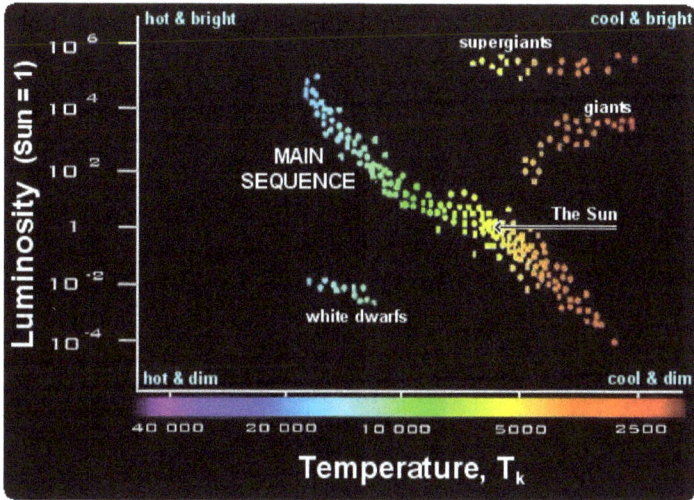

Figure 1.4   The H–R diagram

at a roughly constant temperature because it is a very diffuse body and transparent to radiation, so the heat generated by its collapse is almost immediately radiated away. Eventually, as the density increases, it becomes more opaque to radiation so that as it collapses it starts to heat up. The pressure generated by the increase in temperature slows down the collapse until the collapse comes to a virtual halt. At this stage of its existence, when it radiates in the visible part of the spectrum but has not reached the main sequence, it is known as a YSO, standing for Young Stellar Object, and its slowly-collapsing path towards the main sequence on the H–R diagram can be theoretically derived. These paths for different masses of stars do not intersect each other and this means that, for a YSO, in a position on the H–R diagram to the right of the main-sequence line, we can deduce both its mass and age. I hope this hasn't bored you Sol. It was all to convince you that we can recognize young stars when we observe them and can say how old they are.

Solomon: No Simon, I wasn't bored. It's always interesting to learn the tricks of the trade in other areas of science. You know, it's quite amazing, we just see stars as points of light in the sky, and they can be thousands of light years away, yet we can find out so much about them.

Simon: I know, I give an evening class on astronomy to people without a scientific background and the students seem fascinated by that aspect of the subject. I was once told a story by Professor Round, who was also giving talks to an evening class. After one of his talks an elderly lady came up to him and said 'Professor Round, I think I understand what you were telling us about finding the masses, and so on, of stars but what I don't understand is how we get to know their names!' Anyway back to business.

The way that the intensity of the radiation coming from any hot body varies with wavelength depends on its temperature so in this way we can recognize the temperature of a young star. Here are the Planck radiation curves for some typical stellar temperatures, 5,000, 6,500 and 8,000 K (Figure 1.5). Now what we find is that for many young stars in addition to the curve corresponding to the temperature of the star there is a small bump on the extreme right-hand side corresponding to radiation by a very low-temperature source, a few tens of degrees kelvin. Since low-temperature

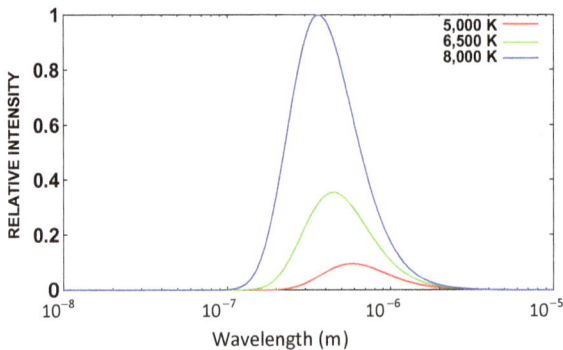

Figure 1.5   Planck radiation curves for three temperatures

sources are very poor radiators and since the bump is high enough to be visible then the source must have a very large area. Everyone, even including Steve, accepts that these bumps are the signatures of extensive disks surrounding young stars and estimates of the masses of these disks vary from one-hundredth to one-tenth of the mass of the Sun.

Steven: Yes I agree, there is no other reasonable explanation but — oh, sorry, I nearly made an interruption but I'll make my remark later.

Simon: You see the point of all this, Sol. For all the problems with solving the angular momentum problem that eventually scuppered Laplace's theory, and no doubt Steve will have something to say about that, we now have direct evidence of disks around YSOs that will eventually evolve to become stars like the Sun. So, theoretical difficulties aside, I want to take this observational foundation as my starting point to describe how planets form in the disk. The angular-momentum problem may be a difficult one to explain but that there is an explanation there can be no doubt.

Steven: This seems like a good time for a short break. My round.

Steven picks up the empty glasses and goes to the bar. He returns 10 minutes later with the glasses refilled.

Solomon: Right Simon, You've done all the groundwork and we accept that some young stars have circumstellar disks. Now tell us how planets form in the disk.

Simon: Essentially it is a four-stage process and I will go through them one-by-one. The first stage involves the dust contained in the disk settling down through the action of gravitational forces to form a comparatively thin dust carpet in the mean plane of the disk. The early work describing this process was carried out by the American planetary scientist, Stuart Weidenschilling and his colleagues and an early problem was that with interstellar dust particles about 0.1 micron in diameter, which execute Brownian motion under constant buffeting by gas molecules, the settling time would be very long — of order many millions of years.[2] An early suggested solution to this problem was that, by one or both of two possible processes, tiny dust grains would stick together to form larger grains that would settle more quickly. One possible process was by 'cold welding'; although we talk of star-forming clouds as 'dense', in terrestrial terms they would be designated as regions of ultra-high vacuum. Consequently the surfaces of grains are virtually uncontaminated and when two grains touch their lattice structures combine to give a single lattice, which is the cold-welding process. The other way that grains can stick together is through electrostatic forces due to particles becoming charged by friction — the kind of forces that make cling-film stick to your hand. Weidenschilling and his colleagues did some computer simulations of how collections of dust particles would assemble and this is an output from one of their simulations (Figure 1.6). Weidenschilling's calculations showed that the settling time was very sensitive to the exact form of the dust aggregates and a shuttle space experiment threw doubt on the simulation results. The experiment called COsmic Dust AGgregation (CODAG) showed that dust particles did adhere but formed fluffy strings rather than blobs. The settling time for these would not be much faster than for the component particles. However, it was thought, but not shown, that if the strings became very long, as they might do after very long periods of time, then they would curl up into a ball-like form.

Much of this early work on dust aggregation has been superseded by recent research, looking at the scattering of radiation by the dust within star-forming regions,

---

[2]Weidenschilling, S.J., Donn, B. and Meakin, P. (1989) *The Origin and Evolution of Planetary Systems*, Eds. H.A. Weaver and L. Danley (Cambridge University Press, Cambridge).

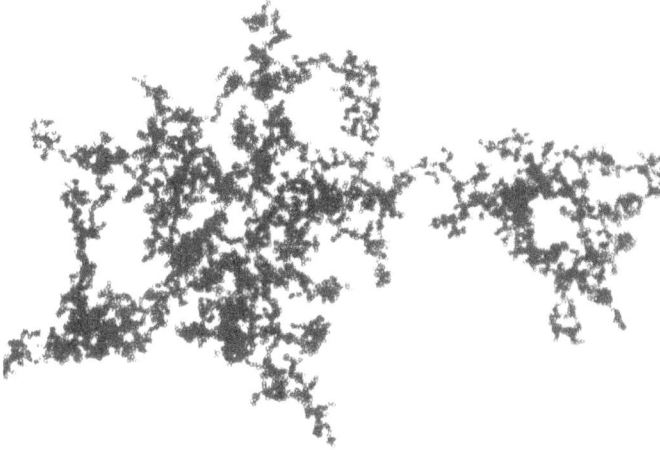

Figure 1.6   A simulation of the aggregation of dust particles

which has shown that some grains are much larger than interstellar grains — up to millimetre size.[3] Such particles would settle on quite a short timescale, perhaps of order hundreds to thousands of years and on their way towards the mean plane of the disk they would sweep up all smaller particles in their paths.

I think it is fair to say that the formation of a dust carpet is generally accepted as a process that could and would readily and quickly occur.

Steven: I'll go along with that. It was once a potential problem for the Nebula Theory but it seems to have been resolved.

Solomon: So far so good. What is the next process?

Simon: The dust carpet would vary in areal density — the mass per unit area — steadily falling with distance from the star but it would not be stable in the form of a smooth continuous sheet. In 1917, James Jeans gave the theory for the instability of a gaseous filament. He showed that

---

[3]Schnee, S. *et al.* (2014) *Monthly Notices of the Royal Astronomical Society*, **444**, 2303–2312.

it would break up into a string of blobs each of which, if it had a great enough mass, would collapse to form a condensed object.[4] In 1973, the American planetary scientists Goldreich and Ward showed that a smooth dust carpet would also be unstable and break up into a number of condensations distributed over the area of the sheet.[5] Following a notation introduced by some earlier workers in the planet-formation field, they called these condensations *planetesimals.*

I'd just like to introduce a few numbers so you'll have a better idea of what the system would look like for the Solar System at this stage. Let us take a disk with mass 10^{29} kg, which is one-twentieth of the solar mass. Since only about 1 percent of that would be dust the mass of dust would be 10^{27} kg, half the mass of Jupiter. Now take the radius of the disk as 35 au; which would comfortably accommodate the orbit of Neptune, the outermost planet, which has an orbital radius of 30 au. In this case the average areal density of the dust disk would be about $12\,\mathrm{kg/m^2}$. Of course, it would be much greater than that close to the Sun and much less in the outer regions. From their analysis, Goldreich and Ward estimated that planetesimals in the Earth region of a disk would have a radius of 1.6 km and radius 42 km in the region of Jupiter. If the average density of a planetesimal was $4{,}000\,\mathrm{kg\ m^{-3}}$ — somewhere between the density of silicates and that of iron — then the mass of a Jupiter-region planetesimal would be about 10^{18}kg. Taking this as an average for the whole system the total number of planetesimals would be 10^9 – one billion. You must not take these numbers too literally but they just indicate that there would be a vast number of planetesimals produced by the break-up of the dust carpet.

---

[4] Jeans, J.H. (1917) *Monthly Notices of the Royal Astronomical Society,* **77**, 186–199.
[5] Goldreich, P. and Ward, W.R. (1973) *Astrophysical Journal,* **183**, 1051–1061.

Steven: You may accept this theory as pretty solidly-based, Sol. What Simon hasn't mentioned is that the process for producing planetesimals is only about 10,000 years, which is very little in the context of the total formation time of the system. However, now we are coming to the more critical bit.

Solomon: What do you mean by 'more critical'? I would have thought that all stages are critical.

Steven: Sorry Sol, I shouldn't have said that, but it will become clearer later on.

Solomon: Carry on Simon. Now explain the third stage.

Simon: To summarize, this part of the process describes how planetesimals come together to form the solid terrestrial planets — Mercury, Venus, Earth and Mars — in the inner part of the system and the solid cores of the giant planets — Jupiter, Saturn, Uranus and Neptune — further out. The theory for this was developed by the Russian planetary scientist, Victor Safronov (1917–1999). It was not developed in the first place in relation to the Nebula Theory but as part of another theory of planet formation due to the Soviet scientist, Otto Schmidt (1891–1956), the Director of the Institute for Theoretical Geophysics in Moscow. In 1944, Schmidt had proposed that the Sun had passed through a dusty dense gas cloud and on its passage through the cloud it had captured some of the cloud material. Safronov, who eventually took over from Schmidt as Director of the Institute, developed theory to explain how this captured material could form planets.

The basic process, as described by Safronov,[6] involved the aggregation of planetesimals by collision. The disk of planetesimals had to be in close-to-free orbit around the central star so that they would come together with a small approach speed. If the approach was too fast then

---

[6]Safronov, V.S. (1972) *Evolution of the Protoplanetary Cloud and Formation of the Earth and Planets* (Israel Program for Scientific Translation, Jerusalem).

the planetesimals might shatter or just bounce apart. The greater the mass of an aggregation of planetesimals the more it was able gravitationally to attract other planetesimals and so grew faster. This led to one aggregation of planetesimals becoming dominant in a number of well-separated regions of the disk, finally giving a few bodies with masses appropriate to those of terrestrial planets or the cores of major planets. Between Mars and Jupiter there is a gap in the natural progression of planetary distances and it seems possible that the large mass of Jupiter, if it formed early, could have scattered planetesimals in the gap region and prevented them from aggregating. This disturbance of planetesimals by Jupiter might also have resulted in the small mass of Mars, which is about one-ninth of the mass of the Earth.

I will not say any more about the accumulation of planetesimals at present. I have described the basic mechanism, the objective of the present exercise, and I know that Steve is going to raise some issues that will enable me to elaborate further.

Solomon: I feel that I would like to know more about this stage but, as you say, there will be an opportunity to talk about it later. Now tell us about the fourth and final stage of planet formation.

Simon: You have to understand that the cores of the major planets, particularly of Jupiter and Saturn, form only a small part of their total mass. For example, the total mass of Jupiter is about 318 Earth masses but the core may only be about 10 Earth masses. The rest is gas, mostly hydrogen but with 10 percent or so of helium and smaller quantities of other gasses. Once you have formed a core within the gaseous disk then it will attract disk gas and any gas impinging on the growing planet will lose energy in the collision and will be retained. This will be an accelerating process, due to the increasing mass of the growing planet, and the accretion of gas will only cease once the gaseous disk has dispersed. The acquisition of a massive atmosphere would be a comparatively quick process, estimated to take about 100,000 years for Jupiter.

That completes my description of the basic mechanism for forming planets. I have mostly mentioned solar-system planets because we know so much about them, but the processes described would also be applicable to the formation of exoplanets, planets around other stars, which have been detected in recent years.

Solomon: Thank you Simon. I now have a pretty good idea about how planets form according to the Nebula Theory. Obviously, this will have to be discussed further since I am sure that Steve will raise some issues with the theory, although what they will be I cannot imagine. I suggest that next week Steve should describe the basic mechanism for producing planets by the Capture Theory. That will give a good basis to explore both models further and to find out the pros and cons for each of them. Is that OK with you, Steve?

Steven: That's fine. I have most of the material I need to hand. You will see that the Capture Theory is quite different from what you have heard today.

Solomon: Good, so let's just finish off our drinks, relax and talk about other things and find out about the basic capture-theory model next week.

They talk for half-an-hour or so and then leave The Anchor.

The Second Meeting

Steven Describes the Basic Mechanisms
of the Capture Theory

It is 5th February 2016 and Steven and Solomon are in the same
secluded cubicle as they were the previous week. Simon approaches
carrying three pints of real ale and sits next to Steven, facing
Solomon. They have a brief chat about events of the past week and
then Solomon brings them to the business at hand.

Solomon: Last week Simon told us about the basic process by which
planets are formed from a nebula. He also gave us an
historical introduction to the Nebula Theory, which I found
helpful since it gave context to the work. You, Steve,
told us that the capture-theory process does things quite
differently so get started and tell me how. If possible I
would like an historical approach. When I give lectures to
undergraduates on the topic of particle physics I always
start with the history of the subject. They all know
already about CERN and the Large Hadron Collider but
they always find it fascinating that the beginnings of the
subject were on a very small scale with equipment that
was assembled in a laboratory workshop and sat on a
laboratory workbench.

Steven: As it happens I have included quite a lot of history in my
account since it shows how the modern Capture Theory
evolved from work originally done more than a century ago.

After the demise of Laplace's nebula theory in the middle of the 19th century, the next theory of any note was put forward by two Americans, the geologist Thomas Chrowder Chamberlin (1843–1928) and the astronomer Forest Ray Moulton (1872–1952) at the end of the 19th century.[1] They had mistakenly interpreted the blurred images of spiral galaxies, which were all that could be seen with the telescopes of the time, as material being emitted from a single star. Here is a modern image (Figure 2.1(a)). The Sun occasionally emits material in the form of prominences as shown here (Figure 2.1(b)) and they suggested that when a particularly large solar prominence was being produced the tidal effect of a passing star pulled the material out further and left it in orbit round the Sun, giving something that would look like a spiral nebula. Then, without giving any detail, they suggested that the drawn-out hot solar material would first form liquid blobs that subsequently coalesced and solidified to give solid bodies. They called these bodies planetesimals, and proposed that they would eventually collect together to form planets; so it was Chamberlin and Moulton who first

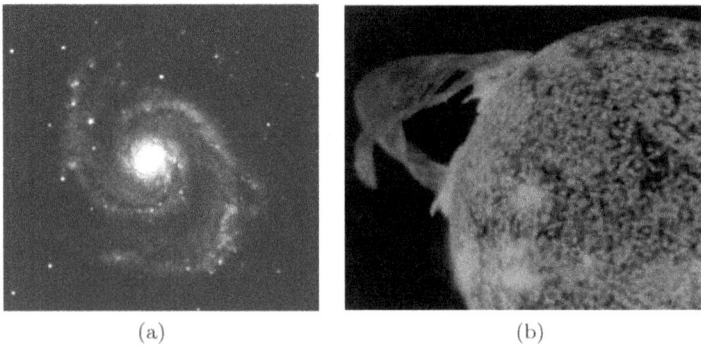

(a)   (b)

Figure 2.1   (a) A spiral nebula, (b) a solar prominence

---

[1] Chamberlin, T.C. and Moulton, F.R. (1900) *Science*, **12**, 201–208.

introduced the term 'planetesimal' that is now used for similar solid bodies in the Nebula Theory, as explained by Simon last week.

A German planetary scientist, Friedrich Nölke (1877–1947), heavily criticized the theory, giving many good reasons why it was untenable, but we don't need to go into those because we now know that spiral nebulae are not single stars but whole galaxies, so the Chamberlin–Moulton theory was founded on a fallacy. However, although the theory was untenable it provided the basis of a scenario for the tidal theory developed by the British physicist and astronomer James Jeans (Figure 2.2).

Nearly all the basic physics associated with the Capture Theory was established 100 years ago by Jeans who developed the theory of a tidal interaction between two bodies, first suggested in a somewhat different form by Chamberlin and Moulton, but only in a rather qualitative way. In 1917 the tidal theory proposed by James Jeans for the formation of solar-system planets became the 'standard

Figure 2.2   James Jeans (1877–1946)

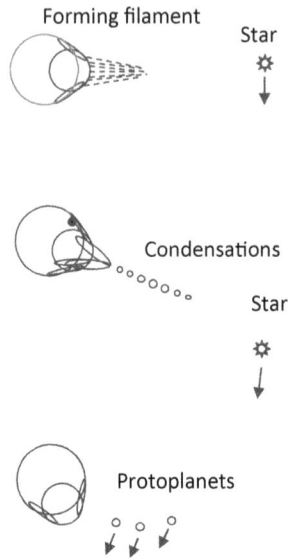

Figure 2.3   The Jeans tidal theory

model' of solar-system formation for almost the next two decades.[2] We can follow the mechanism of the model from this figure (Figure 2.3). A massive star passing close to the Sun drew up a tide so huge that it led to the escape of a filament of solar material. This filament was gravitationally unstable and broke up into a series of blobs, like beads on a string. The blobs were attracted by the retreating massive star and ended up in orbit around the Sun. Each blob collapsed to form a planet. The most massive planets, Jupiter and Saturn, are at the centre of the sequence of planets and Jeans suggested that this was consistent with his model. The filament would have grown thicker as the star approached the Sun and then thinner as it retreated; the thicker central part of the filament would give the most massive planets and planetary masses would tail off both closer to the Sun and further away from it.

[2] Jeans, J.H. (1917) *Memoires of the Royal Astronomical Society*, **652**, 1–48.

The theory is easy to illustrate and hence easy to understand — you did say, Sol, that a picture is worth a thousand words.

Although the theory has been described here in a rather hand-waving way, that was not how it was presented by Jeans. Every component of the model was subjected to detailed analysis. It is really remarkable how the scientists of 100 years ago created simplified models, which needed numerical analysis to investigate, without the benefit of the computers that make comparable work so easy today. Jeans considered how a fluid body, like the Sun, would distort and finally disrupt under increasing tidal forces (Figure 2.4). This figure shows the successive profiles of the Sun as the star approached ever closer. These profiles are what are known as *equipotential surfaces* and Jeans had to calculate these by hand — a complicated calculation that today would be a trivial operation even using the most primitive computer. For the Sun not to disintegrate, the whole of its volume must be contained within a closed equipotential surface. As the star approaches, the equipotential surfaces shrink in scale and the Sun is distorted into an egg shape with the small end getting ever sharper as the star gets closer. Eventually, the small end becomes a point and thereafter no closed equipotential surface has sufficient volume to contain the Sun and material escapes from it through the pointed end in the form of a filament.

Simon has already mentioned the topic of gravitational instability as analysed by Jeans, which he originally applies to a filament, essentially a one-dimensional application, but it also applies in a modified form in two dimensions — as used by Goldreich and Ward for the fragmentation of a dust carpet — and also in three dimensions. We can follow the progress of fragmentation of a filament in a descriptive way in this diagram (Figure 2.5). We imagine that there is an enhancement of density at region A; no gas

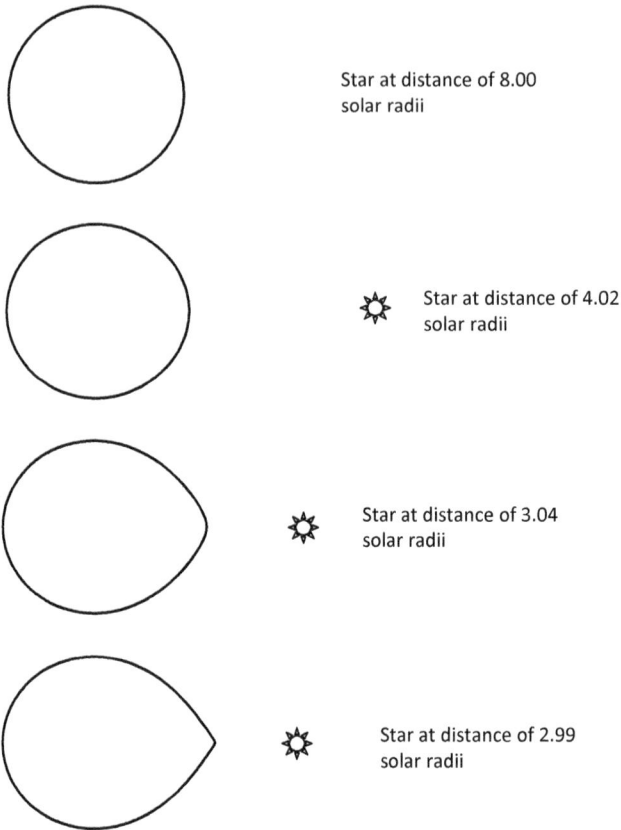

Figure 2.4   Distortion of a solar mass star by one of two solar masses for different separations

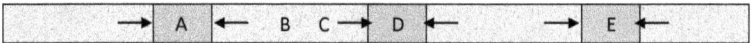

Figure 2.5   The formation of a string of high-density regions due to a density enhancement at A

stream can be completely uniform. Gas on either side of A is attracted towards A with the attraction strongest for the gas closest to A. This creates a lower density in a region such as B so that gas at C will move away from B and produce a higher density at D. Now the high-density region at D acts like that at A and creates another higher density region at E and so on throughout the stream on both sides of A. This wasn't the way that Jeans' described what would happen — his way was more formal and mathematical — and he derived a formula expressed in terms of the properties of the filament for the distance between the condensations.

Finally, Jeans considered the conditions that would enable one of the blobs in the filament to collapse. There are two types of force acting on a blob, gravitational forces that would pull inwards and tend to produce collapse, and gas pressure forces, dependent on the temperature of the material, which push outwards and tend to make a body expand. For a given type and density of material at a given temperature there is a certain mass of the blob — the Jeans critical mass — such that if the blob has a mass greater than the critical mass it will collapse, otherwise it will expand. For a given material of density ρ and temperature T the critical mass is proportional to $\sqrt{T^3/\rho}$.

These analyses were quite rigorous and helped to give credibility to Jeans' tidal theory that initially had strong support; it was generally believed that the problem of the formation of solar-system planets had finally been solved. But this acceptance was not to last. The first objection was raised by the American astrophysicist Henry Norris Russell (he of the H–R diagram) who showed by a back-of-envelope calculation that material could not have been propelled far enough from the Sun to produce Mercury, the innermost planet, let alone those further out.[3] This was essentially another form of the angular momentum problem, in this case because the planets could not orbit at a sufficient distance to acquire the necessary angular momentum. The second major objection was by another

[3] Russell, H.N. (1935) *The Solar System and Its Origin* (Macmillan, New York).

American astronomer, Lyman Spitzer (1914–1997), who used Jeans' own theory to show that, for the density and temperature of material pulled out of the Sun, even a Jupiter mass would have been well below the Jeans critical mass so that Jupiter and the other planets could not form.[4] Eventually the tidal theory was abandoned, but what it left behind was a body of sound theory that has, ever since, been applied in many astronomical contexts. Jeans was an objective and honest scientist, not defending his theory against convincing evidence that it was wrong, and stated 'The theory is beset with difficulties and in some respects seems to be definitely unsatisfactory.'

Solomon: I feel quite sorry for Jeans. He produced all that good theory in support of his model of solar-system formation and, in the end, it came to nothing.

Steven: Don't feel too sorry for him. He was a very successful scientist who made a large number of important contributions to theoretical astrophysics and was one of the leading theoretical astronomers of his time. This idea, that if you follow a possible approach to a problem and it is shown to be untenable you have failed, is wrong in my opinion. There may be several approaches to a problem that may initially seem plausible and examining one of them and showing it does not work may be an essential step to eventually finding the right or, at least, the most plausible, approach. Very often, successful theories contain traces of less successful theories that preceded them. In fact, there is another reason for rejecting the Jeans tidal theory that was not known at the time. The spectrum of the Sun shows that there are virtually no light elements such as lithium, beryllium and boron present in its outer regions. The reason for this is that at solar temperatures these light elements are removed by nuclear reactions. However, the Earth contains appreciable amounts of these

---

[4]Spitzer, L. (1939) *Astrophysical Journal*, **90**, 675–688.

elements so we may be certain that it, and presumably the other planets, were not derived from material at solar temperatures.

Now back to the Capture Theory. In 1964, Michael Woolfson published a paper in which he described a tidal interaction between the Sun and a protostar, a diffuse ball of dusty gas collapsing on its way to becoming a YSO and eventually a main-sequence star.[5] He carried out a simulation in which the protostar was represented by an arrangement of point masses with inter-point forces that simulated gravity and approximately the effects of gas pressure. Because of the limited computer power available at that time he could only represent the protostar in two dimensions with very few points. He then solved the equations of motion for each of the points under the forces to which each was subjected. This figure shows the results he obtained (Figure 2.6). The scale of this model was dictated by the present scale of the Solar System so the radius of the protostar was 14.67 au and the closest approach of its centre to the Sun was 40 au. The protostar can be seen to distort, just as Jeans theory indicated, and eventually a filament is seen to be forming. The model was far too crude to show the gravitational instability of the filament but the paths of many of the points in the filament region showed that they were captured by the Sun (hence the name of the theory) in elliptical orbits with perihelia — closest distances to the Sun — within the range of solar-system planetary orbital radii.

The Capture Theory avoided the drawbacks of the Jeans tidal theory. Material could be in orbit at large distances from the Sun and the protostar material forming the planets was cold, so light elements could be present in the planets. However, what was not shown was that blobs of planetary mass could form and condense. Another

---

[5] Woolfson, M.M. (1964) *Proceedings of Royal Society*, **282**, 485–507.

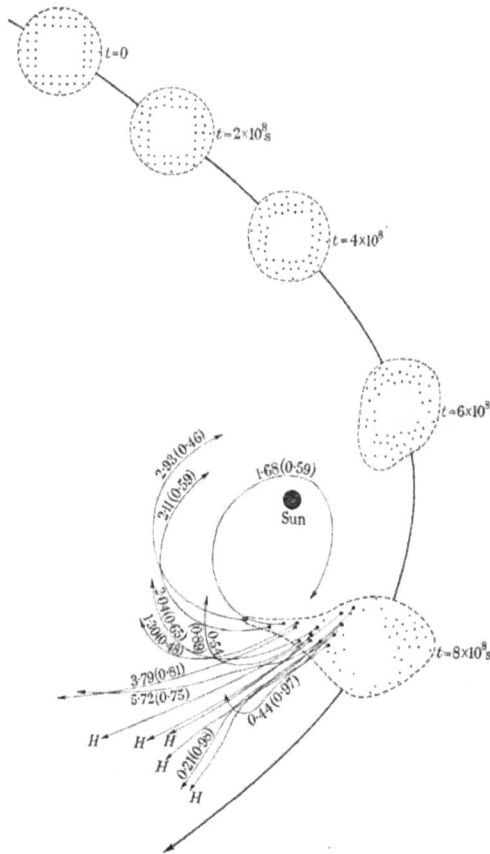

Figure 2.6   The 1964 simulation. Points marked H are on hyperbolic orbits and would not be captured (Woolfson, 1964)

problem, not known in 1964, was that planetary systems are common and the chances of a suitable interaction, either in the present environment of the Sun, where the average separation of stars is about 7 light years, or even in a galactic cluster of stars with separations of order 1 light year, would be very small. We'll come back to the question of probability at a later time. My present brief is just to explain the basic mechanism of planet formation.

Various ways of improving capture-theory simulations were found during the next few years but in 1995 the detection of the first exoplanet — a planet around a distant main-sequence star — was announced[6] and by the beginning of the 21st century it was clear that planetary systems were not rare and might even be described as common. As previously stated, an assumption in the 1964 capture-theory modelling, dictated by the requirement to place the whole event within a region of solar-system dimension, was to take the protostar with radius 14.67 au. Protostars begin their existence as recognizable objects with radii of more than 1,000 au, possibly up to a few thousand au according to some estimates. Such objects would undergo 'free-fall collapse' under gravitational forces, another process first analysed by Jeans. The form of the collapse is that it is extremely slow at first but accelerates and the final rate of collapse to a highly condensed body is very rapid. For example, a uniform protostar with a mass 0.3 that of the Sun and a radius of 2,000 au would take 28,900 years to completely collapse but after one-half of that time it would still have more than 0.8 of its original radius. Hence, for the assumption in the 1964 paper that the protostar had a radius of 14.67 au, the protostar would be in a state of rapid collapse; this was not allowed for in the simulation and, in any case, the state of having a small radius would be of such short duration that the probability of an interaction of a star with a protostar in such a state would be extremely small. It was estimated that just one star in 10^5 or 10^6 might acquire planetary companions in this way.

Solomon: I find all this thinking and learning thirsty work. Give me your glasses and I'll get them refilled.

Solomon takes the glasses and reappears 10 minutes later. After sampling his refilled glass Steven continues.

---

[6] Mayor, M. and Queloz, D. (1995) *Nature*, **378**, 355–359.

Steven: These probability considerations and the detection of many exoplanets meant that the 1964 model was untenable and this led Stephen Oxley and Michael Woolfson to consider an alternative scenario in which the protostar had a large radius, several hundred au or more, and the periastron — a closest approach, like perihelion except applying to a general star — was also very large. By this time, simulations could be done much more realistically. For one thing, computers had improved both in capacity and speed and new computational tools had been designed that simulated the various physical processes that occur in astrophysics extremely well. One such computational technique was smoothed-particle hydrodynamics (SPH), which was similar in type to the point-mass model that Woolfson used in 1964 but much more sophisticated. Each particle had associated mass and specific internal energy — heat energy per unit mass that indicated its temperature — and gravity, pressure and viscous effects were all realistically simulated. Many computational experiments covering a wide range of conditions, showed that the capture mechanism was very robust, working over a large range of scales. Oxley and Woolfson produced a paper in 2004 that showed planet formation and capture, but the simulation gave a mode of behaviour not predicted by Jeans.[7] This figure shows a projected view of their simulation (Figure 2.7). The star has solar mass and the protostar 0.35 solar mass. The radius of the protostar is 800 au and the periastron distance of the orbit of the protostar is 600 au, less than the radius of the protostar. However, no collision of the protostar with the star occurs because almost the whole protostar is drawn into a dense filament, which is gravitationally unstable, just as Jeans showed analytically, and breaks up into a string of blobs. It is pleasing to

---

[7] Oxley, S. and Woolfson, M.M. (2004) *Monthly Notices of the Royal Astronomical Society*, **348**, 1135–1140.

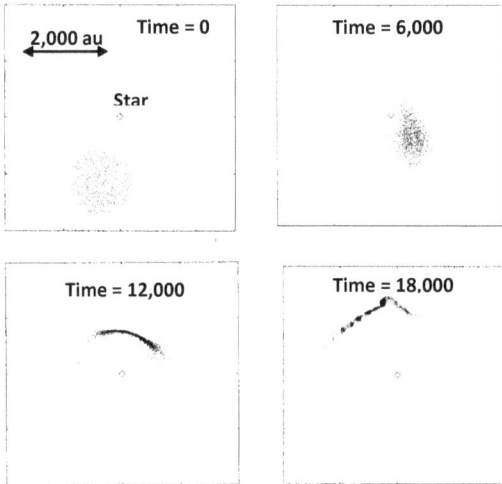

Figure 2.7 A sequence of profiles for a capture-theory SPH simulation at 6,000 year intervals (Oxley and Woolfson, 2004)

see that something that Jeans predicted analytically is reproduced so well computationally. Some of the blobs are not captured by the star but escape into space and these can explain the existence of *free-floating planets*, bodies of planetary mass that have been detected by many observers in interstellar space.[8,9] The remaining blobs are captured by the star, not into near-circular orbits but into wide-ranging elliptical orbits of high eccentricity like this (Figure 2.8). The size and shape of an elliptical orbit are defined by its semi-major axis and eccentricity. In this elliptical orbit around a star the furthest distance, Q, is the apastron and the nearest distance, q, the periastron. The semi-major axis, a, is $(Q+q)/2$ defines the size of the ellipse and the eccentricity, e, is $(Q-q)/(Q+q)$ defines its shape. A circle is a special ellipse with a equal to the radius of the circle and with $e = 0$. The orbits of

[8]Lucas, P. and Roche, P. (2000) *Monthly Notices of the Royal Astronomical Society*, **314**, 858–864.
[9]Sumi, Y. *et al.* (2011) *Nature*, **473**, 349–352.

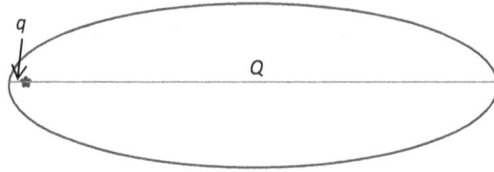

Figure 2.8   The apastron, Q, and periastron, q, of an elliptical orbit

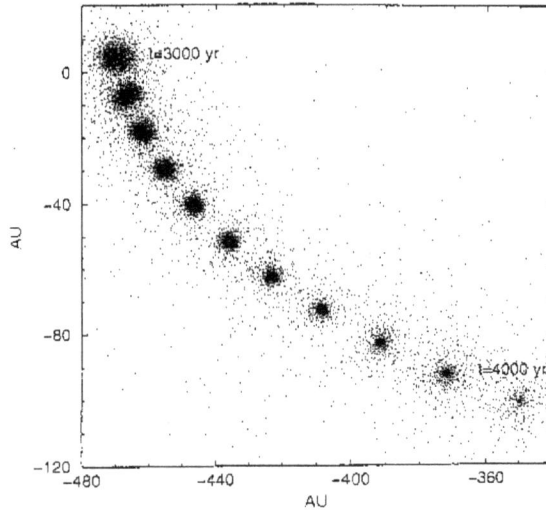

Figure 2.9   The collapse of a protoplanetary blob (Oxley and Woolfson, 2004)

the captured blobs typically have $a = 1,000 - 2,000$ au and $e = 0.90 - 0.95$, very unlike present day solar-system planets and most exoplanets. Obviously I'll have to deal with this question of orbits later but, for now, I am concentrating on the process of forming planets.

The question that couldn't be answered with the 1964 model is whether blobs would form with masses greater than the Jeans critical mass and so collapse to form planets. With an SPH model, in which the blobs contained many particles, that question could be answered. Here we see the collapse of a blob from a capture-theory simulation (Figure 2.9). Over a period of 4,000 years, the blob substantially collapses and would continue to collapse beyond that time. What can also be seen is that the collapse leaves behind a disk of material, which is in the equatorial plane of the spinning planet.

What has been demonstrated is that the tidal effect of a star on a protoplanet can produce planets orbiting the star. In their 2004 paper, Oxley and Woolfson went further. Star-forming clouds are very turbulent with turbulent streams of gas moving at up to 20 km/h, as measured by Doppler shifts of the characteristic radiation they emit, which is in the microwave region of the electromagnetic spectrum.[10] Colliding streams of gas give higher density regions, which initially heat up but cool quite rapidly due to radiation from dust grains and due to atomic and molecular processes. I'll explain the cooling process in more detail in some future presentation. Under favourable circumstances such enhanced-density cool regions can exceed the Jeans critical mass to form a protostar and eventually a normal main-sequence star. However, whatever the potential outcome of a gas-stream collision, if the high density region is produced near a star then the capture-theory mechanism can produce captured planets, as seen here (Figure 2.10), again showing its robustness.

The two simulations I have shown produce captured planets with masses between 0.75 and 20.5 times the mass of Jupiter. The higher end of this range corresponds to bodies called *brown dwarfs*, intermediate in mass between planets and stars, with masses between 13 and 70 times the mass of Jupiter. They generate internal temperatures that give nuclear reactions involving the hydrogen isotope, deuterium, but not high enough to give the fusion of hydrogen nuclei to give helium, which is the prime characteristic of main-sequence stars.

[10]Cook, A.H. (1977) *Celestial Masers* (Cambridge University Press, Cambridge).

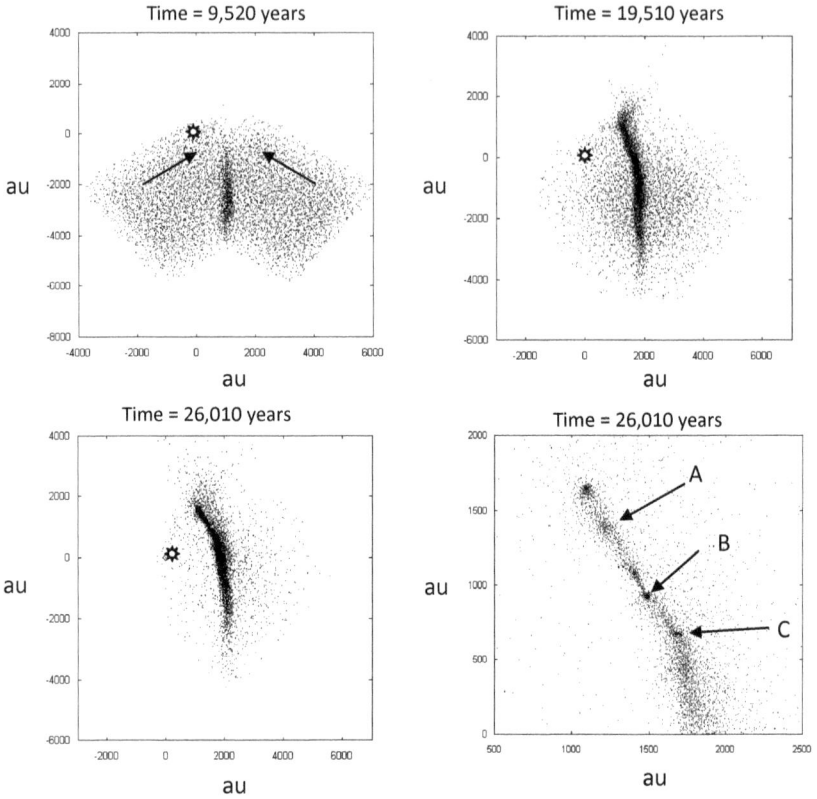

Figure 2.10   Simulation of the interaction of a star with a high density region produced by colliding turbulent gas streams at 3 times — 9,520 years, 19,510 years and 26,010 years. The final image shows a higher resolution view of part of the filament at 26,010 years. The condensations A, B and C are captured (Oxley and Woolfson, 2004)

> I think that is as far as I need to go to show that under a variety of conditions the Capture Theory can give planets around stars — although I accept that there are many loose ends that will have to be discussed later.

Solomon: Well, Steve, you've taken us on a very detailed trip through the intricacies of the Capture Theory and I find the simulations very convincing. For the next session, or perhaps more if it is needed, I would like to hear what you both have to say about the orbits of the planets. Clearly, something has to happen to initial capture-theory orbits to get orbits that agree with what we observe. I am not sure that the Nebula Theory has any such problems but, if it has, I'm sure that Steve will tell us what they are. For now, let us enjoy our drinks and relax.

Steven: I suggest that Simon should be the first to talk and to tell us how the planetary orbits come about through a nebula-theory approach.

Simon: Yes, I am happy with that.

They chat awhile and then leave.

The Third Meeting

Simon is Questioned about Angular Momentum and Planetary Orbits

It is 12th February 2016. Simon and Solomon are in the same cubicle as last week, in a quiet corner of The Anchor. Steven approaches with three pints of real ale and, after sampling the brew, they settle down to renew their discussion.

Solomon: I've been mulling over what you told us two weeks ago, Simon, and it all seems pretty logical and straightforward. However, despite the fact that you justified ducking the issue of the distribution of angular momentum between star and planets by pointing out that disks are detected around young stars, I am still curious to know if any theories have been given for how that observed system can come about. After all, explaining how things that are observed came into being is one of the goals of science. You could hardly say that you are not interested in how planets form because we know they exist.

Simon: Before disks had been detected the whole rationale of going back to a nebula model was partly because of meteorite observations but also because it was believed that with new knowledge the angular momentum problem could be solved. Even before the investigation of new nebula ideas had been fully launched, in 1960 the well-known British astrophysicist, Fred Hoyle (1915–2001), had considered the role that magnetic fields could have in transferring angular

37

momentum.[1] The Sun and similar stars all generate large magnetic fields and Hoyle considered a scenario in which there was a gap between a collapsing stellar condensation and a surrounding disk with the material at the edge of the disk so hot that it would be ionized. The ionized disk material, consisting of positively-charged ions and negatively-charged electrons would be electrically conducting and magnetic fields would become frozen in to this conducting material, i.e. it would act as though it was fixed to the material so that if the material moved it would drag the contained magnetic field with it. This figure (Figure 3.1) shows schematically the central collapsing star and the disk with magnetic field lines linking the two. Now, as the central core collapsed then, to conserve angular momentum, it spun faster. Since the magnetic field lines are rigidly attached to both the core and the edge of the disk they would be stretched. Stretching magnetic

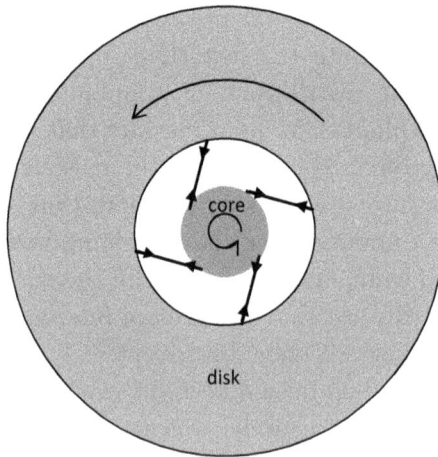

Figure 3.1   Flux lines linking a collapsing core to a disk

---

[1] Hoyle, F. (1960) *On the Origin of the Solar Nebula, Quarterly Journal of the Royal Astronomical Society*, **1**, 28–55.

field lines increases the energy associated with them so, since systems always tend to change towards a minimum energy, the field lines pull inwards at both ends in order to become shorter. This has the effect of pulling backwards on the peripheral material of the core, thus slowing down its spin, and forward on the disk, causing it to spin faster. The overall effect of these effects is to transfer angular momentum outwards, which is just what is required to explain the distribution of angular momentum in the Solar System and, probably, in other planetary systems.

A problem with this theory is that it requires an exceptionally large magnetic field to operate effectively. With the present solar field the field lines would be stretched so far before they exerted an appreciable force on the core and disk that, in order to go to a lower-energy configuration, instead of shortening by pulling inwards on each end, the field lines would simply break and reform in a shorter configuration.

There have been other proposed schemes that use magnetic fields but before describing one of the more effective ones I shall describe a way of transferring angular momentum outwards dependent on purely mechanical forces.

In 1974 the Cambridge-based astrophysicists, Donald Lynden-Bell and James Pringle, considered what would be the outcome if a system consisting of a collapsing core and a connected disk lost energy in some way.[2] This could be because of internal viscous forces within the disk or by material falling on the disk from outside, both of which would convert mechanical energy into heat energy that would then be radiated away. It is fairly straightforward to show that in any isolated spinning system that loses energy, but must conserve angular momentum because it is

---

[2]Lynden-Bell, D. and Pringle, J.E. (1974) *Monthly Notices of the Royal Astronomical Society*, **108**, 603–637.

isolated, the net result is that inner material moves inwards and outer material moves outwards.[3] I can show this very simply (Simon produces pencil and paper to write down equations as he gives his explanation). Consider two bodies in circular orbits, with radii r_1 and r_2 around a central body of much greater mass. The energy, E, and angular momentum, J, of the system are

$$E = -C\left(\frac{1}{r_1} + \frac{1}{r_2}\right)$$

and

$$J = K\left(\sqrt{r_1} + \sqrt{r_2}\right),$$

where C and K are two positive constants. For small changes in r_1 and r_2, the changes in E and J are:

$$\delta E = C\left(\frac{1}{r_1^2}\delta r_1 + \frac{1}{r_2^2}\delta r_1\right) \tag{3.1}$$

and

$$\delta J = \frac{1}{2}K\left(\frac{1}{\sqrt{r_1}}\delta r_1 + \frac{1}{\sqrt{r_2}}\delta r_2\right). \tag{3.2}$$

If angular momentum remains constant then, from this equation (Simon points to (3.2))

$$\delta r_1 = -\sqrt{\frac{r_1}{r_2}}\delta r_2 \tag{3.3}$$

and substituting in this equation (Simon indicates (3.1)) gives

$$\delta E = \frac{C}{\sqrt{r_2}}\left(\frac{1}{r_2^{3/2}} - \frac{1}{r_1^{3/2}}\right)\delta r_2. \tag{3.4}$$

---

[3] Cole, G.H.A. and Woolfson, M.M. (2013) *Planetary Science: The Science of Planets Around Stars* (Second Edition), (CRC Press, Boca Raton).

Given that δE is negative then it can be seen that if $r_2 < r_1$ then δr_2 must be negative, that is to say that the inner body moves inwards and hence, from this Eq. (3.3) the outer body moves outwards. This corresponds to an outward transfer of angular momentum. While this has been demonstrated for just two bodies it is also true for a general distribution of material — in particular a core with a surrounding disk. Inner material will move further inwards and outer material outwards. This process may come into play and create the gap before magnetic forces take over.

In another magnetic-transfer model, in 1996, Armitage and Clarke[4] took the collapsing core to be a highly active pre-main-sequence star with a magnetic field 1,000 times stronger that of the Sun. Magnetic field lines were connected to a wide area of the disk, not just the edge, as shown in this figure (Figure 3.2). Flux lines linked to other slowly rotating outer disk material transfer angular momentum outwards, just as Hoyle described except that with a strong field the flux lines stretched very little. The inner part of the disk, rotating more rapidly, has a different form of behaviour. It is assumed to become rather turbulent and material from this part of the disk

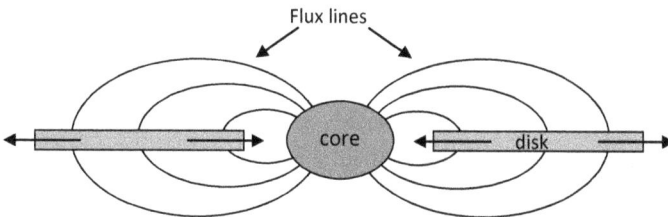

Figure 3.2 Inner disk material moves inwards to join the core while outer disk material moves outwards

[4]Armitage, C.J. and Clarke, P.J. (1996) *Monthly Notices of the Royal Astronomical Society*, **280**, 458–468.

flows along the flux lines to join the central core. Armitage and Clarke showed that this model gave a very efficient outwards transfer of angular momentum.

It isn't necessary for the angular momentum transfer to initially give the Sun, or other stars, spinning at their present rates. They could be spinning much faster, as long as they were rotationally stable, and subsequently lose sufficient angular momentum to explain their present spin rates without being connected to a disk. This process involves the interaction between a stellar wind and the stellar magnetic field. A stellar wind, like the solar wind, consists of a stream of energetic charged particles, mostly protons and electrons, streaming out of the star at a typical speed of $1,000 \, \text{km s}^{-1}$. For the Sun the present solar wind gives a mass loss of 2 million tonnes per second, which may seem colossal but if maintained over the lifetime of the Sun corresponds to a loss of one ten-thousandth of its mass. The escaping charged particles become coupled to magnetic flux lines, moving in a helical path as shown in this figure (Figure 3.3). As the magnetic field weakens with distance from the star, the radius of the helix increases and eventually the charged particle detaches itself from the flux line. Flux lines rotate at the same rate as the star spins and since the coupled particles are increasing their distance from the star but keeping their rotation rate constant then they are gaining angular momentum. Given that the total angular momentum of the system must remain constant the gain of the particles is balanced by loss of angular momentum of the star and its spin rate falls. With a strong magnetic field, as may be expected for an active young star,

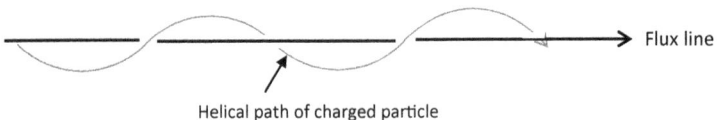

Figure 3.3   The helical path of a charged particle coupled to a flux line

and a very strong stellar wind, observed in some young stars in the so-called T-Tauri stages of their existence, a star spinning tens of times faster than the Sun could lose virtually all its angular momentum while losing less than 10 percent of its mass.

To summarize: I cannot claim that there is a definitive description of how the angular momentum problem is solved, but a combination of mechanical and magnetic processes is probably involved.

Solomon: You have certainly described a range of mechanisms that indicate how a solution to the problem may be found but, as you say, since disks are detected around stars there must be a solution. What do you think, Steve?

Steven: I think that the nebular-theory people are on a completely wrong track. Let us start by assuming that the Lynden-Bell and Pringle mechanism has been working. Although angular momentum will be transferred outwards the inner core of material will always be spinning at such a rate that equatorial material will be in free orbit around the central mass — i.e. it will be on the verge of rotational disruption. When Armitage and Clarke began their analysis for magnetically transferring angular momentum, the starting point for their core was with a tiny fraction of the spin rate that the Lynden-Bell and Pringle mechanism would give. So, how could the Armitage and Clarke initial state come about?

Solomon: Any answer to that one, Simon?

Simon: No, but this is still an area where work has to be done and I am pretty confident that there will be a solution in due course — although the impetus for finding a solution has been somewhat reduced because observations show that disks actually occur.

Steven: I think that I can come to the rescue of the nebula theorists by proposing a solution to the angular momentum problem that gives just what the Nebula Theory needs — don't look surprised, Simon, I am not a convert to the

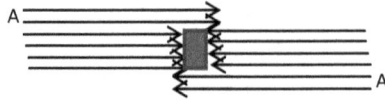

Figure 3.4   The collision of turbulent gas streams. The shaded region represents compressed material

theory — my critical judgement is intact. I pointed out last week that star-forming clouds are turbulent and that collisions between turbulent streams of material would give compressed regions that could give first protostars and eventually main-sequence stars. Now consider what happens when two turbulent streams collide. If one converts the view to one with respect to a stationary centre of mass then it would look like this (Steven draws Figure 3.4). In general, the streams will be offset and dissimilar but the central regions will collide and form a compressed region, which might eventually form a star. However, although most of the peripheral material from the regions marked A would be moving too fast to be captured by the compressed region some of it could be captured and have sufficient mass and angular momentum to provide the disk needed by the Nebula Theory. Of course, this is just a hand-waving description and the scenario needs to be properly modelled to see if it will work; I think it could be done with SPH.

Solomon: We seem to be agreed that, one way or another, we can explain the presence of disks around stars. This apparently establishes the Nebula Theory on a sound basis so what I would now like Simon to do is describe the orbits that would be expected with nebula-theory planet formation.

Steven: Hold your horses, Sol, there is another important problem that Simon has not mentioned. From observations of young stars for which ages can be deduced, as explained by Simon two weeks ago, the lifetimes of disks, before they disperse, is found to be of order 3 million years on average with about 10 million years as a maximum lifetime. This implies that a gas giant planet must form its core well within that time otherwise there will be no disk gas around for it to complete its formation.

I am going to accept that the process of producing a dust carpet and its fragmentation to give planetesimals will be reasonably quick — taking tens or, perhaps, hundreds of thousands of years. That brings us to Safronov's model for the aggregation of planetesimals to form terrestrial planets or the cores of major planets. (Steven now takes more paper and a small calculator from his brief case and puts the paper on the table so that both Simon and Solomon can see what he writes.) Safronov gave an equation for the time it would take to produce a core that can be transformed into this form for a core of mass m and density ρ

$$t = C\frac{P}{\sigma}(m\rho^2)^{1/3}$$

in which P is the period of a circular orbit of a body at the distance from the star where the core is forming, σ is the areal density of planetesimals in the region and C is a constant roughly equal to 0.03. Last week Simon told us that the average areal density of the dust disk was $12\,\mathrm{kg\,m^{-2}}$ and when the dust transforms into planetesimals this will also give us the average value of σ. There are different disk models that are possible but, assuming that the density is highest close to the Sun and falls off exponentially one model gives σ in the vicinity of the Earth, Jupiter and Neptune as about 400, 20 and $2\,\mathrm{kg}$ $\mathrm{m^{-2}}$, respectively. We can also take the average density of a planet core as $4{,}000\,\mathrm{kg\,m^{-3}}$, which is roughly the density of the Earth if we remove the effect of compression due to self-gravitational forces.

If we apply this to the Earth then m is $6.0 \times 10^{24}\,\mathrm{kg}$, and P is 1 year. Putting this into our equation we find

$$t_{\mathrm{Earth}} = 0.03 \times \frac{1}{400} \times \{6.0 \times 10^{24} \times (4{,}000)^2\}^{1/3}$$

$$= 3.4 \times 10^6 \ \text{years}.$$

This time is within the lifetime of many disks and poses no problems. However, for Jupiter with a core mass of, say, $m = 6 \times 10^{25}$ kg and $P = 11.86$ years the formation time comes to 1.7×10^9 years and for Neptune with core mass, say, $m = 3.0 \times 10^{25}$ kg and $P = 164.8$ years the time is 2.0×10^{11} years, much greater than the age of the Solar System. Now you can play around all you like with the parameters for Jupiter and Neptune — for example, you can increase σ by a factor of 10 or even more — but you cannot reasonably reduce the core-formation times to 10^7 years or less to bring them within the observed lifetime of disks. The planetesimal-accretion model, at least as described by Safronov, just doesn't work.

Solomon: That seems to be a pretty strong argument against the Nebula Theory but is the Safronov mechanism generally accepted or, as Steve might be suggesting when he said 'at least as described by Safranov', is it possible that there is another and better mechanism for planetesimal aggregation?

Steven: It seems to be generally accepted although various people have tweaked it here and there to try to reduce the formation times.

Simon: The problem that Steve has just pointed out is well known and there are a number of approaches that can help to solve it.

Solomon: I look forward to hearing what they are.

Simon: Before that let me get these glasses refilled. I find these sessions thirsty work, especially as I have been doing most of the talking!

Simon takes the empty glasses and returns after 15 minutes with the fresh supplies of real ale.

Simon: Sorry about the delay, the bar was busy.

Solomon: Right Simon, we are refuelled so crack on.

Simon: When the lifetimes of disks became known, and the core-formation time problem became apparent, the American planetary scientist, George Wetherill (1925–2006), considered a number of conditions that would speed up the Safronov process.[5]

In the original theory, one growing collection of planetesimals would gradually become dominant in any region and, once that happened, it would fairly quickly sweep up everything else in its region. Establishing this condition, where one aggregation became dominant, could take a considerable time but could be speeded up if initially there was one planetesimal that was dominant, which might arise if the disk were not uniform but rather clumpy in nature. A large clump would give a large planetesimal and a locally dominant body could be present straight away and quickly grow so steadily increasing its dominance.

Next, Wetherill considered the general physics principle 'equipartition of energy'. Under this principle, for a collection of interacting bodies the expected kinetic energy of all bodies will be the same, regardless of mass. Since kinetic energy is given by $\frac{1}{2}mv^2$ it is clear that the larger is the mass, m, the smaller is the probable speed, v. Thus, of all the bodies, the dominant aggregation of planetesimals will tend to move most slowly and of the others the larger ones will tend to have the smaller speeds. This implies that the *relative* speed of a body with the dominant body will tend to be lowest for larger bodies. The range of attraction of one body for another increases as the relative speed decreases, as shown in this figure (Figure 3.5). The rapidly moving planetesimal is hardly deflected as it passes by the dominant body but the slowly moving one is deflected inwards and has a greater chance of being accreted. This tendency for the dominant body to accrete the largest of the remaining bodies should increase the rate of growth of the dominant body above that given by Safronov's theory.

---

[5]Wetherill, G.W. and Stewart, G.R. (1989) *Icarus*, **77**, 330–357.

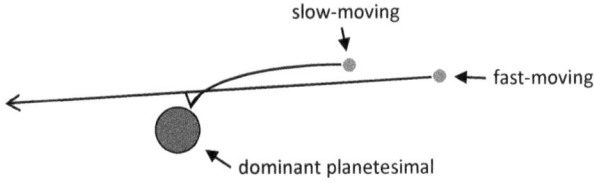

Figure 3.5 The gravitational deflection is greatest for slower moving planetesimals

The next factor considered by Wetherill was gas drag, which will slow down the speeds of all the bodies in the system. This will be more important for small bodies; for bodies of similar shape the drag force will depend on the cross-section of the body so that doubling the dimension of the body increases the force by a factor of four. However, the mass would increase by a factor of eight, so the force per unit mass would be halved. Gas drag will be relatively ineffective in directly slowing down the larger planetesimal aggregations but since the smaller aggregations are losing energy then, through interactions of smaller with larger bodies and the operation of the equipartition principle the speeds of more massive bodies will also be affected. Overall, gas drag may be only marginally important for increasing the growth of the dominant body, but it will have some effect.

Wetherill concluded that adding all these factors could lead to what he called 'runaway growth' that would drastically reduce formation times — down to 1 million years for a Jupiter core and about 30 million years for a Neptune core. The latter time is still uncomfortably large, given the lifetime of disks, but much closer to what is acceptable.

Solomon: It seems that Wetherill has introduced more physics into the process than did Safronov and has made the planetesimal accretion process seem much more viable. Do you have any comments, Steve?

Steven: Yes, I do. Serious doubts have been raised about the runaway model, and these have been expressed not only by opponents of the Nebula Theory but by people who are in the nebula-theory community.

The high rate of growth in the Wetherill model as he and Stewart presented it depends on a very high density of planetesimals in the vicinity of the dominant body. The American planetary scientist, Alan Boss, who generally supports the idea of a nebula origin for planets, has criticized the assumptions in the 1989 paper by Wetherill and Stewart, on the grounds that it assumes a much higher disk density than can reasonably be expected. Another problem was raised by Wetherill himself. A planetesimal striking the dominant body will be accreted by it, but one just passing by will be deflected, the deflection being largest for close passages. The effective scattering cross-section of a body is much larger than the collision cross-section. Once the dominant body reaches about an Earth mass it will be scattering local planetesimals all over the Solar System. The local loss of planetesimals in this way will be partially compensated for by planetesimals being scattered into the region from elsewhere, but these intruders will be moving at high speed and hence will not be readily accreted. In any case, since scattering can be in any direction, including well away from the mean plane of the disk, many more will be lost from any disk region than is gained by it and the local density of planetesimals will fall well below that assumed in the runaway-growth model.

Solomon: That seems to put grave doubts about runaway growth, especially if the author of the theory is also criticizing it. Is there any other salvation for nebula ideas, Simon?

Simon: Yes, there is. Despite the criticism of the runaway-growth model it is still generally believed that some aspects of it will operate and that major planets can form out to about the distance of Jupiter — 5.2 au — within a reasonable timescale. That still leaves the orbits of Saturn, Uranus and Neptune to be accounted for and this has been done by processes that lead to 'planetary migration' whereby planets produced close to the Sun, where formation times are acceptable, move outwards to their present locations.

There are two main types of interaction that can occur, firstly of a planet with the gas in the disk, which can only endure for the lifetime of the gas, and of a planet with residual planetesimals, which can be present round the Sun, or any other star, for hundreds of millions of years after the gaseous part of the disk has dispersed. The underlying physical principle is the same in both types of interaction. If the effect of the planet is such as to add angular momentum to external material then, by the conservation of angular momentum, the planet will lose angular momentum and move inward. Conversely, if the interaction causes external material to lose angular momentum then the planet will gain angular momentum and move outwards. To explain the positions of the outer planets of the Solar System, we are interested in the outward migration of those beyond Jupiter, if we accept that they can form within the lifetime of the gas disk much further in than they are now.

The theory of some of the effects I am going to describe can be rather complicated but I am going to simplify it as much as I can, perhaps missing some of the subtleties of what is going on but getting the essential description correct. We can begin by considering what the effect of a planet is on a particle or a parcel of gas in an orbit exterior to itself. The particle will have a lower orbital speed than the planet and if we look at its motion relative to a frame in which the planet is at rest then as we see here (Figure 3.6) the relative motion of the particle is in the opposite direction to the actual motion.

As the particle is approaching the planet its speed relative to the planet increases and, conversely, as it

Motion of particle
relative to planet

• planet
Actual direction of motion
of planet and particle

To star

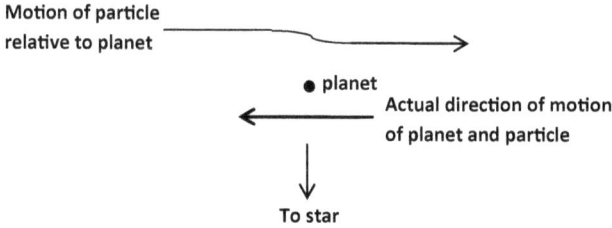

Figure 3.6   The relative motion of particle and planet

retreats from the planet its relative speed decreases. However, as is seen in the figure — somewhat exaggerated for clarity — in its passage past the planet it is attracted towards the planet so the decrease in relative speed slightly exceeds the increase. The net effect is a small decrease in relative speed and this corresponds to an increase in the orbital speed around the star. This gives an increase in the angular momentum of the particle in its orbit and a corresponding decrease in that of the planet. Thus, the planet moves inwards and the particle moves outwards. If we now consider the motion of a particle orbiting inside the planet's orbit then the reverse happens — the planet is pushed outwards and the particle inwards. If there were two particles, one inside and one outside the planet's orbit then, whether the planet would move inwards or outward, would depend on the relative strengths of the inward and outward forces. However, what is certain is that the particles on either side of the planet move outwards, away from the planet. Considering the motion of a planet in a gaseous disk the overall effect is that the planet tends to create a region of lower gas density along its orbital path. Before explaining how this will affect the planet's motion I wish to describe another mode of interaction between a planet and a particle.

As physicists we are all aware of resonance as a phenomenon. A large number of small-energy impulses delivered to a physical system at intervals in synchrony

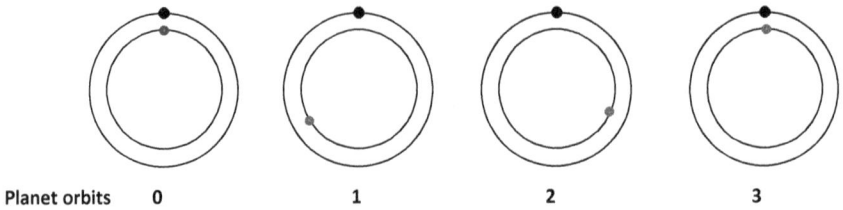

Planet orbits   0   1   2   3

Figure 3.7   The planet and particle positions at intervals of the planetary period

with some natural frequency of the system can build up and steadily increase its energy. An everyday example is that of repeatedly giving a small push to a child's swing when it is at the end of its backward motion. The amplitude of the swing's motion steadily increases. Now let us consider a particle in an orbit interior to a planet such that it makes four orbits while the planet makes three. If we look at the position of the particle relative to the planet at intervals equal to the period of the planet then this diagram (Figure 3.7) shows the positions of the particle.

It can be seen that the closest approach of planet and particle always occur at the same point of the particle orbit and this resonance situation leads to an enhanced effect on the particle orbit and a correspondingly large reaction on the planet in the form of a torque, either increasing or decreasing its angular momentum. This kind of resonance is called a 'Lindblad resonance' and occurs when the ratio of planet to particle periods is "$m:m-1$" for inner particles and "$m:m+1$" for outer particles, where m is an integer.[6] There can also be a co-rotational resonance in which, if the disk gas is not in Keplerian motion, i.e. rotating at the same rate as a solid body at that distance, then the gas can be rotating with the same period as the planet but not in the same orbit.

---

[6]Tanaka, H., Takeuchi, T. and Ward, W.R. (2002) *Astrophysical Journal*, **565**, 1257.

These interactions with the gaseous disk lead to the following types of planet migration. First there is Type I migration, where the planet has insufficient mass to open up a clear passage in the disk. The strongest effects on the planet are due to Lindblad resonances; in the great majority of cases the outer Lindblad resonances, which drives the planet inwards, overcome the inner Lindblad resonances so that the planet moves inwards. However, it is possible to construct a disk model, with a concentration of mass close in to the star, where an outward motion of the planet is produced.

The effectiveness of a planet in clearing a gap in the disk is proportional to the square of its mass. A Jupiter-mass planet would be expected to clear a gap but for one of Saturn mass it is doubtful that a clear gap would be formed. Where there is a clear gap then Type II migration occurs. The gap wipes out most of the close inner and outer Lindblad resonances for large m and the effects of the residual inner and outer resonances tend to cancel. Because of viscosity due to gas at different distance from the star moving at different speeds, the inner part of the disk tends to lose energy and move inwards. The planet, and the gap, move inwards following the inner part of the disks. The rate of migration inwards is on a 10^5 year timescale and this kind of motion probably explains the existence of giant exoplanets, of Jupiter mass or greater — so-called 'hot Jupiters' — very close to their stars, in positions where the high temperature would inhibit the formation of an atmosphere.

Solomon: I must say that so far you have made a much stronger case for inward migration than outward migration, although you have said that for some disk structures outward migration is possible.

Steven: While you are on the subject of hot Jupiters perhaps you should explain to Sol why it is that they do not plunge into the star.

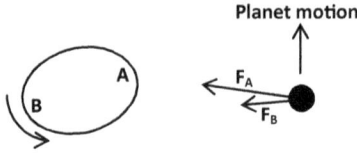

Figure 3.8   A star tidally distorted by a planet

Simon: Yes, I'll do that before describing the migration mechanism due to interactions of planets with planetesimals. I haven't pre-prepared a visual aid for this explanation so if you'll give me a sheet of your paper, Steve, I'll do a quick diagram (Simon sketches out Figure 3.8).

Here we show a tide raised on the star due to a massive planet; the scale of the tide has been exaggerated for clarity. The spin of the star drags the tides forward in the direction of the spin and if the orbital period of the planet is greater than the spin period of the star then the position will be as shown in the figure. Now we can approximate the force of the star on the planet as that due to a spherical body, which acts from planet-centre to star-centre, plus those due to the two tidal bulges, $\mathbf{F_A}$ and $\mathbf{F_B}$. Because bulge A is closer to the planet, $\mathbf{F_A}$ is larger than $\mathbf{F_B}$ and $\mathbf{F_A}$ has a component along the direction of the planet's motion. This corresponds to an increase in the angular momentum of the planet, which would cause it to move further from the star. However, this is balanced by the opposite tendency of the Type II mechanism to reduce the angular momentum so the planet is stabilized in its orbit. It is quite stable; if the planet were to move closer to the star then the tidal force would increase and dominate and push it out again. On the other hand, if it moved further from the star then the migration mechanism would dominate and it would move inwards. An observable example of this kind of tidal effect is seen in the Moon's slow retreat from the Earth. In this case, the tidal effect is operating but there is no migration mechanism to oppose it so the Moon gradually moves further from the Earth.

Solomon: That was interesting and explains those hot-Jupiters close to stars you mentioned. Look, I've had a lot to digest tonight and I suspect that the planet–planetesimal interaction story may not be a short one so I suggest that we break off now and carry on with it next week.

Simon: It isn't too long or too complicated. Depending on Steve's reaction to it I reckon that it will take half an hour or so.

Solomon: In that case, Steve, after Simon completes what he has to say would you be ready to tell us about how you reconcile the very extensive planetary orbits produced by the Capture Theory with what is now observed.

Steve: That's fine. I'll have all my stuff ready.

They sit for a few minutes discussing the detection of gravitational waves which had been announced the previous day. Then they finished off their drinks and departed.

The Fourth Meeting

Simon Describes Planet–Planetesimal Interactions and Steven Explains How Initial Capture-Theory Orbits are Modified

It is 19ᵗʰ February 2016. Steven and Simon are in their usual seats and Simon is looking through some papers. Solomon approaches, accompanied by the landlord of The Anchor carrying three glasses of real ale on a tray.

Landlord: Good evening gentlemen. I have noticed that you have become Friday regulars, arriving by 8 o'clock, always occupying this rather quiet cubicle and engaging in intense conversation. Your friend here tells me that you are discussing an interesting scientific problem, almost certainly something that would be way over my head. Tonight you were lucky; this cubicle was occupied until a few minutes before you arrived. Would you like me to put a reserved notice on the table next week — that is if you are coming next week.

Solomon: Thank you landlord, we are certainly coming next week and we would be most grateful if you would reserve this cubicle for us.

Landlord: Consider it done, gentlemen. Have a good discussion. (*He departs.*)

Solomon: That was nice of him. Now to business — off you go, Simon. Tell us about the role of planetesimals in producing outward migration.

Simon: One of the difficulties I have here is that a large amount of work has been done in this area and some of the analyses in the resulting papers are complex and are very difficult to describe in words; a Web search will reveal the large number of approaches that have been made and the processes they involve. However, the basic idea behind planet–planetesimal interactions is pretty straightforward and obvious. If the planet–planetesimal interaction causes the planet to gain angular momentum then it will move outwards — just a little at each interaction because planets are so much more massive than planetesimals. Conversely, if the interaction causes the planet to lose angular momentum then it will move inwards. On the face of it you might think that one kind of interaction is just as likely as the other and whether the net effect on a planet is that it moves inwards or outwards is just a matter of chance.

I am going to give my own simplified version of what could have given dominantly outward migration. The Nebula Theory would have tended to give planetesimals initially moving in near-circular orbits, following the motions of the dusty gas from which they were derived. Again, planets moving within a gaseous resisting medium that the nebula disk provided would also have had near-circular orbits. Now in this simple diagram (Figure 4.1) in (a) I show the result of an interaction of an interior planetesimal with a planet in the frame of the planet at rest. Relative to the planet the planetesimal is moving in the same direction as the planet in its orbit because its orbital speed is greater, and it is thrown outwards by the interaction. Throughout its motion the planetesimal will be exerting a gravitational force on the planet and from the diagram it is evident that the addition of velocity

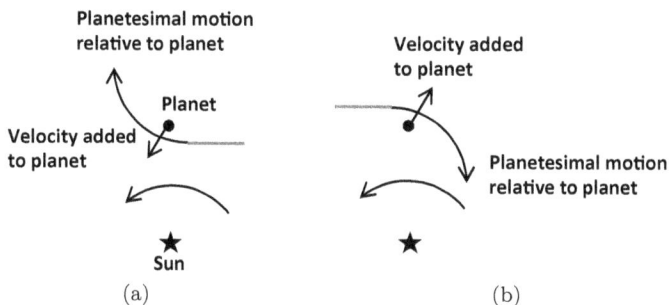

Figure 4.1   Interaction with a planet with (a) an inner planetesimal and (b) an outer planetesimal

to the planet has a component in the direction of its motion so the angular momentum of the planet will be increased. In (b) an exterior planetesimal, has a relative velocity in the opposite direction to the planet's motion. It is thrown inwards but the additional velocity added to the planet has a component in the opposite direction to its motion so its angular momentum is reduced. In both cases, the orbit of the planet will become more eccentric. This shows that for a net outward movement of the planet we need events such as (a) to happen more often than events such as (b).

Assuming that the total mass of the disk is 0.1 solar mass and that 1 percent of this is dust, the total mass of dust, and hence approximately the mass of planetesimals, would have been about the mass of Jupiter, 2×10^{27} kg. Although the early areal density of planetesimal material was in proportion to the areal density of the gas in the disk, due to Type I migration involving the interaction of planetesimals and gas, which is typically inwards and rapid on a few times 10^5 year timescale, the planetesimals would have quickly migrated inwards giving a high density of planetesimals close to the Sun together with a very high density gradient with density decreasing outwards. This high areal density of planetesimals close in would have decreased the time of formation of terrestrial planets and giant-planet cores so helping to solve that problem. Although up to one-tenth of a Jupiter mass of planetesimals would have been taken up in producing terrestrial planets and major-planet cores there would still have been a density gradient of planetesimals decreasing outwards and for that reason interactions of type (a) in my figure should dominate over those of type (b) and the planet should migrate outwards. Although this approach is simple, and no analysis has been given, I hope that I have indicated the plausibility of the outward migration of planets by their interactions with planetesimals.

Solomon: Yes, I accept that you have shown the plausibility of outward migration but it would need some kind of theoretical or numerical analysis to make it completely convincing. What do you think, Steve?

Steven: Simon's description is fine, as far as it goes, but there are many issues it does not address. The inward migration of planetesimals due to gas drag is certainly helpful for planetesimal aggregation close to the Sun or, in general, to a star but increasing the areal density of planetesimals at the centre of the system is tantamount to denuding the outer part of the system. Neptune has to migrate out to a distance of 30 au, Uranus to 19 au and Saturn to 10 au. Would the angular momentum of planetesimals remaining in the outer part of the system be sufficient to produce the migration of these outer planets sufficiently far, even if all their angular momentum were given up to the planets? It would be difficult to find out — the modelling parameters would be hard to define, but it is an area of uncertainty.

Simon: I think that the model I have suggested is possible and I am sure that a set of parameters can be found to give the required result. After all, the Solar System formed somehow and it wasn't constrained to form by processes that can be accurately modelled with certainty. No plausible theory has to satisfy that criterion.

Steven: I'll accept that, but we now have increasing evidence of the existence of exoplanets at distances from stars much greater than 30 au. Most exoplanets have been detected either by Doppler-shift measurements of the wobble of stars due to orbiting planets — with both star and planet orbiting the centre-of-mass of the two-body system — or by the diminution of the intensity of light coming from a star if a planet transits across its disk and blocks some of the light coming from it. By the nature of these observations they only give information about planets close to stars — much closer than Uranus and Neptune are to the Sun. Recently some astronomers in California have claimed that from the motions of objects in the Kuiper Belt, beyond the orbit of Neptune, they have detected the presence of a body they call the 'ninth planet' orbiting the Sun at a distance of about 600 au.[1] Again, nowadays, with improvements in telescope technology, there are direct observations of planets much further from their stars than the observed outer solar-system planets are from the Sun. These observations are of very hot massive planets that strongly emit infrared radiation, by which they are imaged. The first such image was produced by the American astronomer Paul Kalas in 2008 when he showed the image of a planet orbiting the star Fomalhaut. This photo (Figure 4.2) shows how the planet moved between 2004 and 2006. The planet is estimated to have an orbit of semi-major axis 115 au with eccentricity 0.11, corresponding to an orbital period of 872 years. However, the extreme case is a directly imaged planet of estimated mass 11 Jupiter masses at 650 au from the young star

[1]Batygin, K. and Brown, M.E. (2016) *Astronomical Journal*, **151**(2) Article No. 22.

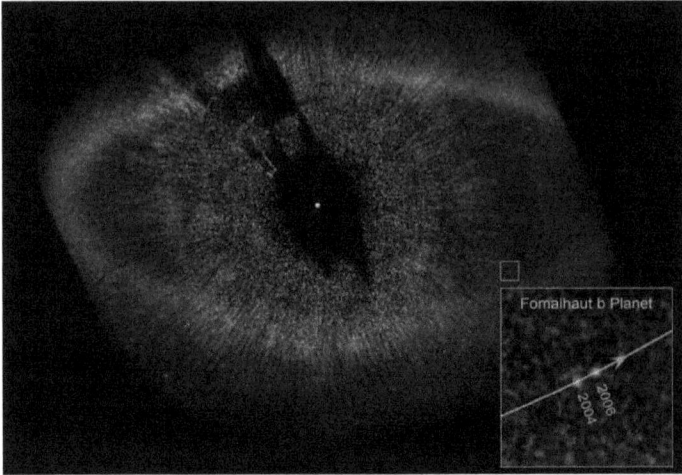

Figure 4.2   The motion of the planet orbiting Fomalhaut between 2004 and 2006

HD 106906, just 12 million years old.[2] The people making these observations of exoplanets distant from their stars have commented that they could not possibly be explained by migration from a formation position much closer to the star. What is your reaction to this, Simon?

Simon: There are many possible ways to explain planets at large distances from stars. The dynamics of the early stages of planet formation are likely to have been very chaotic. Interactions between the planets of a system can result in some of them either being projected to great distances from stars or even escaping from the star altogether.

All the exoplanets we are interested in are associated with main-sequence stars, like the Sun, that begin their existence in a cluster of stars called either *galactic clusters* or sometimes *open clusters*. They contain anything from a few hundred to a few thousand stars and they are produced by the collapse of regions within star-forming clouds. During the period, a cluster of stars is being

---

[2]Bailey, V. *et al.* (2014) *Astrophysics Journal*, **780**, L4.

produced within the collapsing region it is said to be an *embedded cluster* meaning that the stars are embedded in gas. The stars are also collapsing inward with the cloud and eventually the stellar density can become very high so that stars are only separated by a few thousand au. This proximity of stars gives two possible ways of giving exoplanets at great distances from stars. The first is that a close passage of a star past another star with a planetary system can perturb one or more of its planets into more extended orbits. Another possibility is that, in the many-body environment of a dense embedded cluster, a planet escaping from one star can be captured by another, but in an orbit with a large semi-major axis.

Steven: What you say is possible. However, capture requires very special conditions and is somewhat difficult to achieve. Two isolated bodies coming together from a great distance will end up a great distance apart unless there is some mechanism for extracting energy from the two-body system — such as a collision or a very close tidal interaction. The presence of other bodies can remove energy but for capture of a planet by a star the only other bodies likely to be closely present are other planets in orbit around the star but because of their comparatively low mass they would be unable to take up much energy. A far more likely event is perturbation of a planet to a more distant orbit, as you have mentioned, so I would more-or-less rule out capture.

There is one other aspect of exoplanet orbits that you ought to explain, which is the large eccentricity of some exoplanet orbits. The majority of exoplanet orbital eccentricities are fairly small but there are many in the 0.3–0.7 range and one extremely eccentric orbit is that of the planet HD 80606b, with semi-major axis 0.447 au and eccentricity 0.934.

Simon: There is no great mystery about these. The particular planet you mention, HD 80606b has a mass about 4 times that of Jupiter and comes in the category of being a 'hot Jupiter' that we previously mentioned. Now, if a similarly massive planet was also migrating inwards at the same time and made a close interaction with HD 80606b then the other planet could have been expelled from its bound orbit around the star while HD 80606b was sent into its present very eccentric orbit. Since inward migration occurs so readily we might expect that many interactions between migrating planets take place in the region close to a star to give highly eccentric orbits.

Solomon: I think that Simon has described how the orbits of both solar-system planets and exoplanets can be explained — although what we have heard is just indicative of how orbital features can be explained rather than giving a precise theory or convincing simulations of how things happened.

Steven: That is a feature of nebula-theory ideas. Simon mentioned how large and varied was the literature relating to planetesimal–planet interactions. Many take this plethora of published papers as an indication of the vitality of the field but I do not see it that way. The fact that nobody can produce a theory convincing enough to shut the topic down — job done, so to speak — seems to me a sign of weakness, especially when so many of the ideas put forward are very convoluted. In fact, Simon's simple model seems to me to be far better than many of the complicated and esoteric explanations I have seen.

Solomon: Well Steve, you have been very forthright in your criticism of what Simon has been telling us. You finished your explanation of the basic capture-theory mechanism with planets on extended orbits with high eccentricities. Can you do any better in explaining the present orbits of solar-system planets and exoplanets?

Steven: I think I can, but before I do that I think it about time we drained our glasses and refilled them again.

They finish their drinks and Steven goes to the bar, returning a few minutes later with three full glasses.

Steven: Two weeks ago, when I described the basic mechanism for producing planets you saw that a protostar or dense region, produced by a collision of gas streams, was stretched into the form of a very thick and dense filament within which protoplanetary condensations formed. What I did not mention, and was not very evident in the figures I showed, was that some of the protostar or dense-region material — usually between 25 and 50 Jupiter masses of it but sometimes of greater or lesser mass — was captured in the form of a disk around the star. Later on I'll show you another figure that makes this capture of material evident.

To get to the matter in hand I am going to point out a general principle that will help my explanations. There are various ways in which a solid body can interact either with another solid body or with a gas. If you think about the interactions of planets with either the gaseous nebula or with planetesimals, the general rule is that energy is lost by the faster moving body and gained by the slower moving one. Thus, for a planet in a gaseous nebula in a Keplerian orbit the faster inner gas loses energy and moves inwards while the slower moving planet gains energy and moves outwards. This principle also applies to viscosity forces due to motion of a solid body through a fluid. If you move your hand through water you experience a force in the opposite direction to the motion of your hand. The faster your hand moves, the larger is the force: the physics of the situation tells us that the force is proportional to the speed of motion. I should stress that this general principle applies to the scenarios we are considering and may not be of universal application in all cases.

Now I am going to consider the motion of a planet in a highly eccentric orbit in a gaseous medium that is in Keplerian orbit (Figure 4.3). The red arrowed-lines show the speed of the planet both at periastron and apastron.

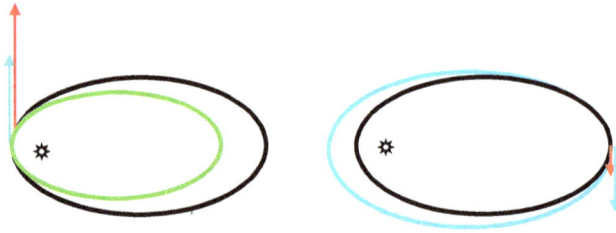

Figure 4.3   The orbital speeds of the planet at periastron and apastron are marked in red. The medium speed at periastron is shown as green and the effect of the different speeds slows down the planet and modifies its orbit to the green form. Similarly the apastron speed of the medium is shown blue and the effect of the different speeds increases the speed of the planet and modifies the planet orbit to the blue form

At periastron, the speed of the gaseous medium is shown in green and is less than that of the planet. This leads to a slowing of the planet and its orbit is changed to the green form, with periastron unchanged but a reduced apastron. This has the effect of reducing both the semi-major axis and the eccentricity of the orbit. On the right-hand part of the figure, the speed of the medium at apastron is shown in blue and is greater than that of the planet. Hence, the planet is speeded up and goes into the blue orbit with apastron unchanged but periastron increased. The effect of this is to increase the semi-major axis but decrease the eccentricity. At both orbit extremes, the effect is to decrease the eccentricity so the overall effect is certainly to round off the orbit. At periastron the semi-major axis decreases and at apastron it increases but since the density of the medium usually decreases outwards the effect at periastron dominates and the net effect is that the orbit decays, i.e. the semi-major axis decreases. Of course the medium is acting on the planet during the whole of its orbit, not just at periastron and apastron, but the nature of the overall effect is given correctly by just considering what is happening in these two regions, as is confirmed by a computational approach.

The computational approach was made by modelling the gas as a distribution of point masses.[3] When reference is made to disks around stars the usual image conjured up is that of a flattish structure of almost uniform thickness, rather like a coin. However, it isn't possible for a gaseous disk to take up such a configuration since it would be unstable. For a stable configuration, the cross-section of the disk would have to flare outwards with distance from the star. For this reason, the distribution of points representing the disk is as shown here (Figure 4.4). The cross-sectional density in the model has been made independent of distance from the mean plane of the disk although it should actually fall off with distance. Another factor that must be included in any model is that the disk will disperse on a timescale of a few million years, as has been observed for disks around young stars. This has been simulated by having the density at any point at time t after the formation of the disk as (Steven writes)

$$\rho(t) = \rho(0)\exp(-\alpha t),$$

where $\rho(0)$ is the initial density and the constant α controls the rate of dispersal of the disk.

In the computation approach, the resisting medium was represented by 77,408 points, as shown in the figure, the masses of which were set to give the required distribution of medium density. These point masses were in motion in Keplerian orbits around the star but they did not gravitationally interact with each other so that in the absence of a planet the medium was stable. When a planet was inserted, its motion was affected by the mass of the star and the masses of the medium points. This figure (Figure 4.5) shows the result of a calculation with the following parameters. The planet was of one Jupiter mass

---

[3] Woolfson, M.M. (2003) *Monthly Notices of the Royal Astronomical Society*, **340**, 43–51.

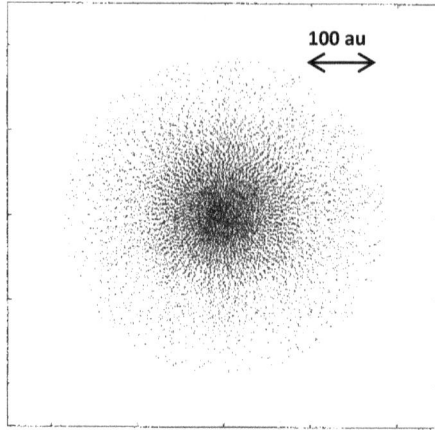

(a)

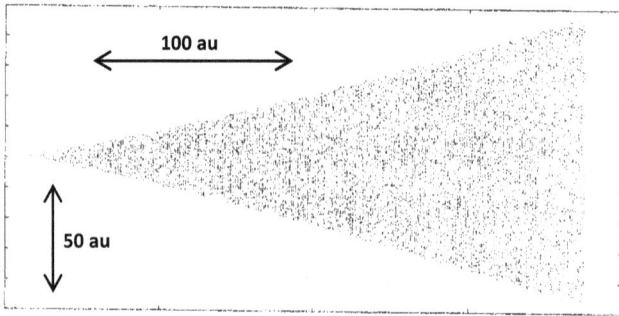

(b)

Figure 4.4   The distribution of particles representing the medium in the mean plane of the disk. (a) and in a cross section of the disk. (b) (Woolfson, 2003)

and had an initial orbit with semi-major axis 2,500 au and eccentricity 0.9. The mass of the medium was 50 Jupiter masses and it consisted of material of temperature 20 K and mean molecular mass 2×10^{-27} kg. The areal density of the material fell of exponentially with distance from the star such that it halved every 139 au and the decay of the medium was such that it halved every 693,000 years. The star had solar mass. The orbit completely rounded off in about 2.7 million years and in 4 million years the semi-major axis stabilized at about 5 au.

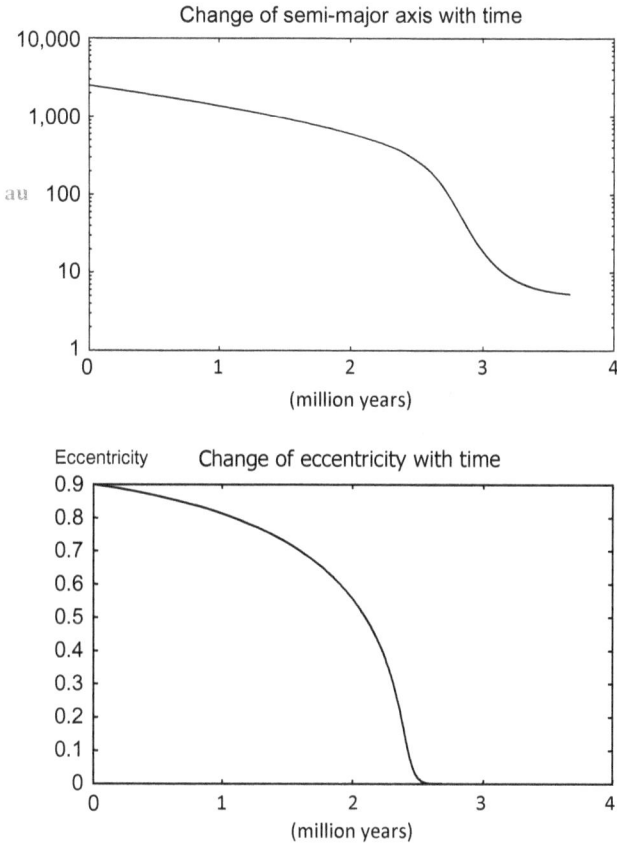

Figure 4.5   Change of semi-major axis and eccentricity with time in a resisting medium

This kind of modelling has been carried out with a large range of parameters. Some of the main conclusions are as follows. When the planetary mass was varied the final outcome always gave round-off but the final semi-major axis was approximately inversely proportional to the mass of the planet. As would be expected, the rate of orbital decay was reduced with a lower medium mass and the extent of round-off and decay was less for a shorter-lived medium. Depending on the mass, distribution and duration of the medium a very large range of final semi-major axes could result. Thus, a massive and long-lasting medium would give the conditions for 'hot Jupiters' while a low mass, low duration medium would give planets stranded at large distance from stars.

With the resisting media in Keplerian orbits, and an areal density that fell exponentially with distance from the star and had a few million year existence before substantially decaying, the outcome was always complete, or nearly complete, round-off. However, the assumption of the gas being in Keplerian orbit is not always valid. Early stars can be very luminous and have very strong stellar winds and these will have a significant effect on the medium although virtually no effect on a planet. It has been estimated that the early Sun may have been 60 times as luminous as it is at present with a solar wind between 10,000 and 100,000 times stronger than now.[4] Radiation and stellar winds push outwards on the medium and neutralize part or all of the gravitational force of the star. Indeed, the effect can be to overpower the effect of gravity and disperse the medium. Now here is a diagram, similar to one I showed you previously, giving the relative speeds of planet and medium when the medium has been greatly slowed down by the effect of a very active star (Figure 4.6). At periastron the effect is as before, although a bit stronger because the medium has slowed and the force on the planet is greater. The periastron is unchanged and the eccentricity is reduced. At apastron, because of the slowing down of the medium, the planet is again slowed down leading to a smaller periastron and an increase in eccentricity. At both extremes, the orbit decays but there are opposite effects on the eccentricity. If the areal density of the medium decreased with distance from the star then the effect at periastron would dominate and the overall

---

[4]Kuhi, L.V. (1964) *Astrophysical Journal*, **140**, 1409–1432.

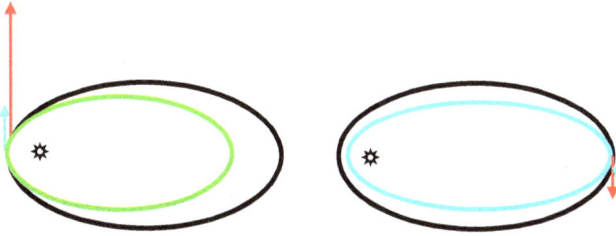

Figure 4.6   The orbital speeds of the planet at periastron and apastron are marked in red. The medium speed, reduced by the stellar wind, is shown at periastron as green and the planet's orbit is modified to the green form. At apastron the medium speed, shown in blue, is so reduced by the stellar wind that it is now less than the planet speed. Hence the planet is slowed by the medium and the planet orbit is modified to the blue form

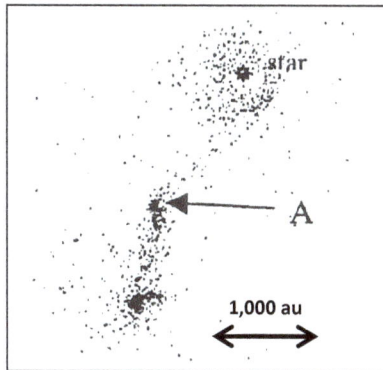

Figure 4.7   A capture-theory simulation showing a strong doughnut-like captured medium (Woolfson, 2003)

result would almost certainly be both decay and round-off of the orbit.

A final point is that in all cases, with initial orbital inclinations relative to the mean plane of the medium up to 20°, the final inclinations were all close to zero.

In the many simulations of the Capture Theory that have been carried out some of them gave a captured medium that did not fall off monotonically with distance from the star but instead took on a doughnut configuration. Here is one example (Figure 4.7). This means that when the orbit has decayed and only partially rounded-off the situation changes so that the density at apastron can be much greater than that at periastron. In that case, the apastron effect, that of increasing the eccentricity, can dominate and the eccentricity can increase. By the way, I told you that I would be showing a clearer example of captured protostar material forming a resisting medium. This is it, with the captured material enhanced to make it more obvious.

To confirm this conclusion a number of calculations were carried out and showed how, with a doughnut medium and a strong stellar wind, the eccentricity at first reduces, when the whole orbit is large and later increases when the apastron is within the peak of the doughnut region. Two of the simulations shown here (Figure 4.8), with large reductions in effective stellar mass due to the stellar wind, give final orbits that are highly eccentric. By varying the parameters of the simulation much higher eccentricities can be produced.

A characteristic of the orbits of the major solar-system planets is that they seem to be related in pairs, giving what are called 'commensurate orbits'. Thus, the orbital periods of Saturn and Jupiter are in the ratio 29.46:11.86 or 2.48:1, which is close to 2.5:1 and that of Neptune to Uranus is 164.8:84.02 or 1.96:1, which is close to 2:1. In 1996, Melita and Woolfson showed that when the orbital periods of a pair of planets, decaying at different rates in a gaseous resisting medium, became commensurate, then they could subsequently become locked together so that while the orbits continued to decay they did so in a way that kept the ratio of orbital periods constant.[5] This figure (Figure 4.9) shows the variation of the ratio of periods of two planets starting at 2.5:1 and both decaying

---

[5]Melita, M.D. and Woolfson, M.M. (1996) *Monthly Notices of the Royal Astronomical Society*, **280**, 854–862.

Semi-major axis (au)

time (million years)

Eccentricity

time (million years)

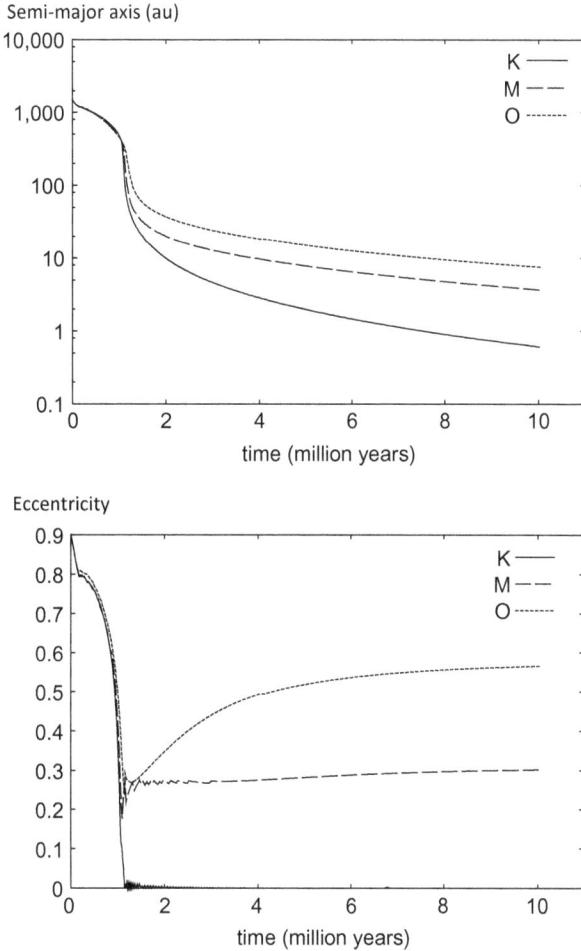

Figure 4.8 Three simulations of orbital evolution, two of which give eccentric orbits (Woolfson, 2003)

in a resisting medium. Nine runs of the simulation are made with different pairs of eccentricities chosen from 0.1, 0.2 and 0.3 for each of them. Four runs settle down to a ratio of periods close to 2.5:1 while the other five settle down to close to 2:1. This shows the effect of the medium in coupling the planets together and producing commensurability quite well.

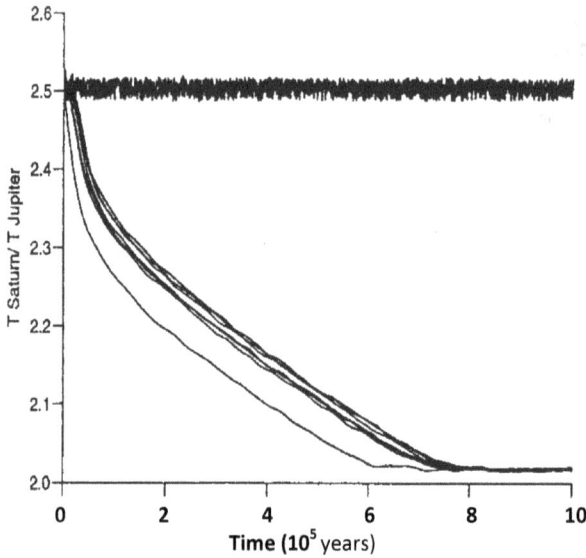

Figure 4.9   Evolution of a Jupiter–Saturn system starting with different eccentricities (After Melita and Woolfson, 1996)

This mechanism depends on the fact that for two bodies in commensurate orbits energy is removed from the inner body by the outer one and added to the outer body by the inner one. For a Jupiter–Saturn pair when a resonance is reached the addition of energy to Saturn partially compensates for the loss of energy due to the resisting medium and the orbital evolution of Saturn is slowed. However, the loss of energy from Jupiter due to Saturn adds to the loss due to the resisting medium and Jupiter's orbital decay is enhanced. This disparity in the rates of orbital decay quickly removes the resonant condition, thus slowing down Jupiter's decay and increasing that of Saturn. Saturn's orbit now decays more quickly, by the Type 1 migration process, than that of Jupiter by the Type II migration process and resonance is soon re-established. You can see how the orbital periods became locked to whatever value they have once resonance is firmly established.

That completes my description of how the initial very extended and eccentric orbits are modified to what we now observe in the Solar System and for exoplanets. Fortunately all aspects of the orbital modifications I have dealt with are amenable to simulation so that one can be reasonably confident that the processes described could actually take place.

Simon: But that does not mean that they would actually occur. If the capture-theory scenario does not take place then, no matter how well orbital evolution can be modelled it just wouldn't happen that way.

Solomon: This has been a long session and my brain is overloaded with new ideas. I'll spend the next week looking at my notes and your figures and sorting it all out. Simon has brought up the topic of how likely the capture-theory scenario would occur, claiming that capture-theory events may not actually occur, so we might ask the same question of the Nebula Theory. Let's talk about that next week and I suggest that you kick off, Steve.

Steven: That's fine with me.

The friends chat for a few minutes while they empty their glasses, then leave The Anchor and go their separate ways.

The Fifth Meeting

Steven and Simon Talk about the Proportion of Stars with Planets

On the 26th February 2016, Solomon and Simon arrive together at The Anchor and find that, true to his word, the landlord has placed a reserved card on their cubicle table. Simon goes to buy some ale and while he was away Steven arrived, a little breathless, saying that the underground train had been stuck in the tunnel for 10 minutes for some unknown reason. Simon returns with the drinks and the friends settle down in their usual places.

Solomon: Well Steve, you have given us a good description of the capture-theory mechanism and of how orbits evolve but both Simon and I remain to be convinced that the capture-theory can explain the thousands of exoplanets that have been detected.

Steven: Actually the latest count is that 2,084 planets have been detected around 1,329 stars with 510 of those stars having two or more planets. If there is one thing we can be sure about it is that a plausible theory of planetary formation must give the conclusion that exoplanets are common and that is what I am going to show for the Capture Theory.

Before going further I would like to define a unit of distance that is usually used by astronomers in dealing with interstellar distances; for the usual separations of stars or distances of stars the astronomical unit is inconveniently small. The way that the distances of very close stars are determined is by means of what is called the *parallax method* was first suggested by Johannes Kepler in the 17th century, although the technology of that time was inadequate to implement it. The basis of the method is shown in this figure (Figure 5.1). The position of the near star is recorded on the background of very distant stars from two points on the Earth's orbit 6 months apart. These two points, A and B, are chosen so that AB is perpendicular to the direction of the star. The projected positions on the background of very distant stars are A′ and B′. The angle α, which is at most about a second of arc, must be found to determine the distance of the near star but what we can actually measure is the angle $\alpha - \varepsilon$ subtended by A′B′ at either of the points A or B. However, since the distances of background stars can be thousands of times greater than that of the near star, ε is very much smaller than α so effectively we measure α. The distance of the star is proportional to the inverse of α, the angle in seconds of arc subtended by the diameter of the Earth's orbit in au at the star's position, i.e. $2/\alpha$. and the distance is then in the units *parsecs*, abbreviated to pc. Thus if α is 2 seconds of arc the distance is 1 pc and if α is 1 second of arc the distance is 2 pc. This unit is used because it is directly related to what is measured — an angle. The parsec is of the same order of size as the light-year, the distance travelled by light in 1 year; 1 pc

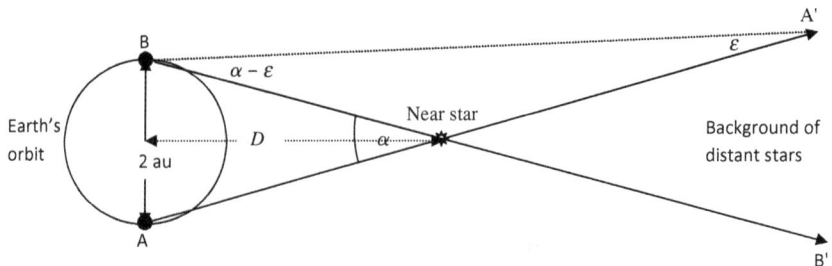

Figure 5.1   Finding the distance of a nearby star by the parallax method

equals 3.262 light-years, which equals 3.086×10^{13} km or 206,265 au.

Last week, when he was giving some ideas about how planets could become very distant from the stars they orbit, Simon mentioned the dense embedded state of a forming galactic cluster. The stars, and the gas in which they are contained, are all collapsing inwards and protostars are being produced while this collapse is taking place. At any time, the latest forming bodies are protostars while earlier formed bodies are one of collapsing protostars, YSOs and main-sequence stars. Eventually, the collapse of the cluster is halted by the internal pressure of the system, which builds up as the collapse proceeds, and for a period of the order of a few million years there is a high number-density of stars within the embedded cluster. The lifetime of stars on the main sequence decrease with increasing mass and is about 10^{10} years for a solar-mass star. However, later in the development of a cluster more massive stars form. Actually, there is a theoretical problem with forming massive stars directly. If you took, say, 10 solar masses of gas that was collapsing to form a star then at some stage the temperature of the forming core would be so high that the radiation from it, pushing outwards, would prevent further material from joining the core. However, stars of 10 solar masses and much more, do form, although they are not very common. The answer to this problem was suggested by the British astronomer Ian Bonnell and his colleagues. In the environment of a dense embedded cluster, whole stars, YSOs or even dense protostars can aggregate to form large mass stars.[1] The importance of massive stars in the context of what I am talking about is that they have a very short lifetime on the main sequence,

---

[1]Bonnell, I.A., Bate, M.R. and Zinnecker, H. (2002) *Monthly Notices of the Royal Astronomical Society*, **298**, 93–102.

at the end of which they explode as supernovae. The lifetime of a star on the main sequence varies as $M^{-2.5}$, where M is the mass, so a star with 20 times the mass of the Sun would have a lifetime of between 5 and 6 million years. The energy generated by supernovae drives out the gas from the dense embedded cluster and with the gravitational influence of the gas removed the cluster begins to expand. In 90 percent of cases the expansion continues indefinitely and the stars or binary systems in the cluster become isolated *field stars* or *field binaries*, isolated entities within the galaxy without nearby companions. In the other 10 percent of cases the end product is a galactic cluster of stars. Whatever the final outcome, the short lifetimes of massive stars on the main sequence limit the period of maximum stellar density to a few million years.

I think I ought to give you some idea of stellar separations in various environments. In the vicinity of the Sun the local stellar number density is about $0.08\,\mathrm{pc}^{-3}$. This corresponds to an average stellar separation of $2.3\,\mathrm{pc}$. In the central region of a fairly dense galactic cluster of stars there may be 500 stars in a spherical volume of $0.5\,\mathrm{pc}$, corresponding to a stellar number density of about $1,000\,\mathrm{pc}^{-3}$ and an average stellar separation of $0.10\,\mathrm{pc}$ or $21,000\,\mathrm{au}$. The stellar number densities within dense embedded clusters have been observed with values up to several times $10^4\,\mathrm{pc}^{-3}$. However, this may not correspond to the maximum density that can occur; the cluster may be evolving towards, or away from, that state. Some authors[2] claim that the maximum stellar number density may be as high as $10^6\,\mathrm{pc}^{-3}$. However, to stay within the realm of observations a stellar number density of $2.5 \times 10^4\,\mathrm{pc}^{-3}$ corresponds to an average stellar separation of about

---

[2] Adams, F.C., Proszkov, E.-M., Faluzzo, M. and Myers, P.C. (2006) *Astrophysical Journal*, **641**, 504–525.

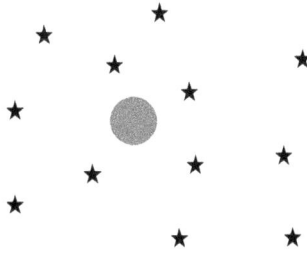

Figure 5.2 A two-dimensional representation of a newly-formed protostar (circle) within a cluster of stars

7,000 au; similar to the 8,000 au separation mentioned by Simon last week.

This two-dimensional figure will help you to envisage a three-dimensional situation where we have newly-formed protostars of diameter anywhere between 2,000 and 4,000 au in an environment where stars are, on average, 7,000 au apart (Figure 5.2).

Given that the star and protostar are in motion with speeds, according to Gaidos,[3] in the range 0.5–2.0 km s^{-1} it seems likely that a proportion of protostars will approach a star close enough for a capture-theory event to take place. The Capture Theory is amenable to a computational approach to investigate the proportion of star that would acquire one or more planets.[4] The various capture-theory simulations that have been carried out indicate that any interaction in which the closest approach to a solar-mass star is between 0.5 and 1.5 times the current radius of the protostar gives planet formation. For a star of mass M_* the limits of distance D, for a capture event are given by (Steven writes)

$$0.5 \left(\frac{M_\odot}{M_*}\right)^{1/3} R_p \leq D \leq 1.5 \left(\frac{M_\odot}{M_*}\right)^{1/3} R_P,$$

[3] Gaidos, E.J. (1995) *Icarus*, **114**, 258–268.
[4] Woolfson, M.M. (2016) *Earth, Moon and Planets*, **117**, 77–91.

in which $M_\odot$ is the solar mass and R_P is the current radius of the protostar.

Since only a finite number of simulations have been carried out it cannot be claimed that it is always true that D within these limits will give a capture event but enough simulations have been done to indicate that the majority of such interactions will give planets. The limits of closest approach to give planets are somewhat fuzzy; it is assumed that closer approaches will lead to break-up of the protostar without planet formation and that for further approaches the protostar will not be drawn into a filament. Nevertheless, for interactions close to but outside both of the given limits planets can sometimes form.

For the computational approach, the initial system of stars is set up in a cubical cell structure, shown in two dimensions in this figure (Figure 5.3). A number of stars, N_S, is placed in randomly-chosen positions in the cell of side a, which is chosen so that N_S/a^3 gives a preselected stellar number density. The same distribution is placed in 26 'ghost cells' (eight in the two-dimensional figure) surrounding the main cell.

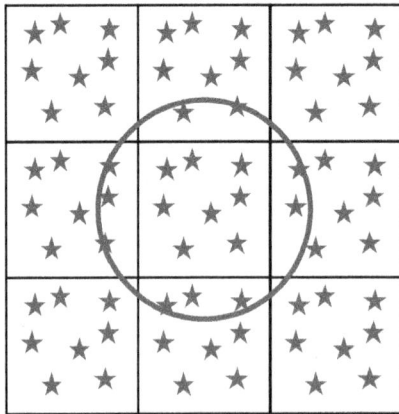

Figure 5.3   A two-dimensional representation of a cell structure with limiting ring of radius ma

A high-resolution SPH simulation by Bonnell and colleagues[5] showed that in the final stages of collapse of a star-forming cloud the cloud first consists of a number of regions containing a few tens of stars in which the stellar number density can be up to $2 \times 10^5 \, \mathrm{pc}^{-3}$ although the average stellar number density in the whole cloud, consisting of many such well-separated regions, is a few times $10^3 \, \mathrm{pc}^{-3}$. The fragments begin both to expand and to come together so that the stellar number density in the fragments decreases but that in the whole cloud increases. In the final stages of the simulation, there were 400 stars in five fragments with a whole cloud stellar number density about $2 \times 10^4 \, \mathrm{pc}^{-3}$ but higher than that in the individual fragments.

The investigation of the frequency of planet formation was carried out in a region corresponding to one of the final fragments, assumed to be of spherical form. The fragment was defined as the region within a distance ma of the centre of the main cell; this boundary is shown as the circle in the figure. This gives the total number of stars as (Steven writes)

$$N_T = \frac{4}{3}\pi m^3 N_S.$$

From observations,[6] and also a theoretical approach,[7] the first stars produced in a forming cluster have masses averaging about 1.3 solar masses. Thereafter, there is one stream of development where masses decrease with time and later another stream develops where masses increase with time. This latter stream we can identify as

[5]Bonnell, I.A., Bate, M.R. and Vine, S.G. (2003) *Monthly Notices of the Royal Astronomical Society*, **343**, 413–418.
[6]Williams, I.P. and Cremin, A.W. (1969) *Monthly Notices of the Royal Astronomical Society*, **144**, 359–373.
[7]Woolfson, M.M. (1979) *Philosophical Transactions of the Royal Society London*, **291**, 219–252.

due to the merging of previously-formed stars as suggested by Bonnell and his colleagues. From this pattern of development, we can deduce that newly-formed protostars are likely to be less massive than pre-existing stars. In the simulation being described here, the masses of the stars were randomly selected between 0.5 and 3.0 solar masses from a observationally-derived distribution (Steven writes)

$$f(M) \propto M^{-2.3},$$

where $f(M)$ is the number of stars per unit mass range. This figure (Figure 5.4) shows the distribution, with the characteristic that low mass stars are the most common. The computational approach was to place a protostar of mass 0.3 solar masses and initial radius R_P within the stellar region, but with the constraint that initially it could not be too close to a star on the grounds that protostars could not form in a strong tidal field. Then the star and protostar were set in motion in random directions with speeds that satisfied the equipartition principle, so that the least massive bodies had the highest speeds.

In an isolated system of interacting bodies — where, for example, the interactions can be by elastic collisions or by

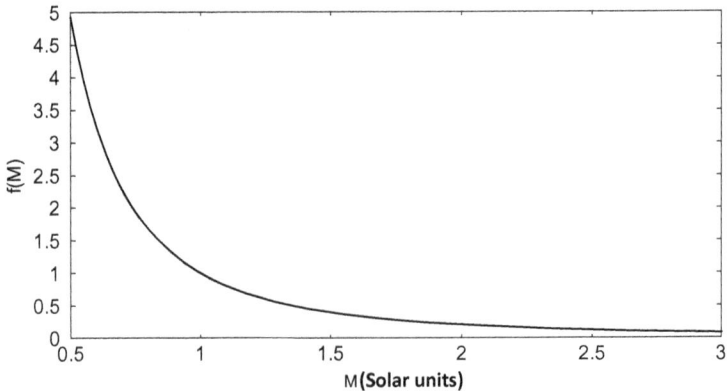

Figure 5.4   The function $f(M)$

gravitational forces — then, in an equilibrium state the kinetic energy of the system, K, is related to the potential energy of the system, Ω by the *Virial Theorem* (Steven writes)

$$K = -0.5\Omega.$$

The gravitational potential energy of the star system is easily calculated but in allocating the speeds account was taken of the slowing of the stars through their motion through gas, which was done by replacing 0.5 by 0.04 in the *Virial Theorem*[8]; This reduced the average speed of the stars, which had the effect of reducing the number of planet-forming interactions.

The calculations were carried out for three values of R_P — 1,000, 1,500 and 2,000 au — for which the free-fall times were 10,200, 18,800 and 28,900 years, respectively. The equations of motion for all the bodies in the central cell were numerically solved in a series of short timesteps. If the protostar approached the star too closely, the trial calculation was abandoned and recorded as too close. If there was a closest approach between 0.5 and 1.5 R_P then a planet-forming trial was recorded and, again, the trial terminated. As the solution proceeded the radius of the protostar was reduced at the free-fall rate but if there were no outcome after half the free-fall time, when the protostar radius was 0.8 of its original value, then the trial was abandoned as having given no capture-theory event. At the end of each timestep if a star, or protostar, moved out of the central cell then it was reintroduced into the cell as though it had passed through the face opposite to the one where it had exited and the ghost cells were modified accordingly; in any event the ghost cells were reset at the end of each timestep. In this way the stellar number density

[8]Protszkov, E.-M., Adams, F.C., Hartmann, I.W. and Tobin, J.J. (2009) *Astrophysical Journal*, **697**, 1020–1032.

Table 5.1   Percentage of capture-theory interactions with various R_P and stellar number density, n

R_p (au)	n	Set A Capture (%)	Set B Capture (%)	Set C Capture (%)	Average Capture (%)	Set D Capture (%)
1,000	5,000	0.7	0.6	0.3	0.5	0.3
	10,000	1.8	1.6	1.0	1.5	0.8
	15,000	3.4	2.1	1.8	2.4	1.9
	20,000	3.8	3.3	2.9	3.3	3.2
	25,000	5.3	4.5	4.6	4.8	5.3
1,500	5,000	2.3	2.3	2.7	2.4	2.4
	10,000	6.5	5.6	5.9	6.0	5.5
	15,000	11.6	8.9	10.2	10.2	11.0
	20,000	11.5	13.5	11.9	12.3	12.9
	25,000	15.9	16.3	17.1	16.4	19.5
2,000	5,000	7.9	6.6	5.0	6.5	5.2
	10,000	17.7	16.9	16.7	17.1	17.0
	15,000	29.0	25.9	24.7	26.5	29.1
	20,000	36.4	37.8	33.6	35.9	40.8
	25,000	48.0	46.0	44.2	46.1	52.4

Note: Set A: $N_S = 8$, $m = 1.5$, $N_T = 113$, mass of gas equals mass of stars.
Set B: $N_S = 23$, $m = 1.0$, $N_T = 96$, mass of gas equals mass of stars.
Set C: $N_S = 10$, $m = 1.0$, $N_T = 42$, mass of gas equals mass of stars.
Set D: $N_S = 10$, $m = 1.0$, $N_T = 113$, mass of gas equals 3× mass of stars.

in the simulated fragment was kept constant. For the three values of R_P, 1,000 trials were carried out for stellar number densities 5,000, 10,000, 15,000, 20,000 and 25,000 pc^{-3}. The mass of gas was also taken into account in terms of its gravitational effect on the motions of the stars and protostar. This table (Table 5.1) shows the percentage of capture events for various combinations of parameters. The results are also shown graphically in this figure (Figure 5.5) where the curves are smoothed to eliminate the statistical fluctuations in the numerical results.

The results do not directly lead to a proportion of stars that gain planets but rather indicate the proportion of protostars that give planets. In general, it would be expected that there would be more protostars than stars so, as a conservative estimate the results in the table can

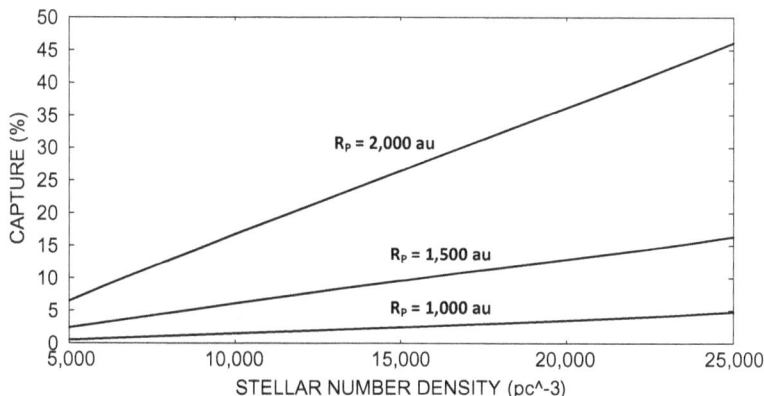

Figure 5.5   Proportion of stars with planets for different protoplanet radii (R_P) and stellar number densities

be taken as the proportion of stars with planets for the given parameters.

The best current observationally-based estimate of the proportion of stars with planets is 0.34.[9] Since the statistics for the maximum stellar number densities in the dense embedded state is not known it cannot be claimed that the figures given in the table certainly explain the observational estimate. However, given the very conservative parameters used and the fact that capture-theory interactions with colliding dense regions are not included, it seems likely that the estimate obtained from observations, or even-higher future estimates could be accommodated.

Solomon: What do you mean by conservative estimates?

Steven: The whole calculation was carried out in a way that, if anything, underestimated the number of planetary systems formed. For example, the stellar number density was restricted to 25,000 pc^{-3}. The Bonnell *et al.* simulation of the evolution of an embedded cluster gave much higher stellar number densities in the early stages. If 10^5 or $10^6 \, pc^{-3}$ had been used, as some workers in the field

---

[9]Boruchi, N.J. *et al.* (2011) *Astrophysical Journal*, **736**, 19–40.

have suggested, then the estimated proportion of stars with planets would have been much higher. Some authors have given much larger initial radii for protostars, which would also give a larger estimated proportion. Again, in the analysis, protostars were not allowed to form too close to existing stars but it has been shown that colliding streams near stars can lead to planet formation, although not via the formation of a protostar. If the analysis had taken non-conservative values of parameters, and pushed the boundaries of what was possible, then it would have been estimated that almost all stars had accompanying planets.

Solomon: Fine — let's have a short break while I get the glasses refilled.

Solomon goes to the bar and returns a few minutes later with a tray of brimming glasses.

Solomon: Now, Simon, what do you think about Steve's presentation?

Simon: I have a few doubts about it. Steve tells us that there are more protostars than stars, yet in his model he only has one protostar. Surely there will be many protostar–protostar interactions. I am not sure of the full implication of that but surely it must reduce the number of capture-theory events.

Solomon: That seems a good argument. What do you say to that, Steve?

Steven: Within the timescale of the development of the stellar cluster, once a YSO is formed then it remains as a condensed object thereafter. On the other hand, protostars are transient objects — we only follow then for half a free-fall time — and although there may be more of them than stars they do not all exist at the same time. There may be more than one protostar present at any one time, or even none, but there will not be so many that protostar-protostar interactions would be common and should be taken into account.

Solomon: I would like to think about that argument but, in the meantime, Simon, what can the Nebula Theory tell us about the proportion of stars hosting planets?

Simon: It isn't possible to carry out an analytical or computational approach to estimate the proportion of stars with planets in relation to the Nebula Theory. However, as I have said previously, there is no condition imposed on actually-occurring astronomical processes that they should be easily amenable to such approaches. Observations suggest that the disks around stars have masses in the range one-hundredth to one-tenth of a solar mass. The solid content of a minimum-mass disk would have a mass of about 50 Earth masses, more than enough to give the terrestrial planets and the cores of major planets and for more massive disks there would be an abundance of solid material. Taking the whole mass of a minimum-mass disk, both gas and dust, it is 7 times the mass of all the solar-system planets. Although I am looking at these disk masses in terms of producing the solar-system planets, the evidence we have suggests that many stars have a much smaller total mass of planetary companions while others have at least one companion with several times the mass of Jupiter. It is reasonable to suppose that there is a direct link between the mass of the planetary material produced and the mass of the disk.

That is about as much as can be said on this topic for the Nebula Theory. A large proportion of young stars are observed to have disks. If the processes invoked by the Nebula Theory are valid then there it is almost inevitable, especially for larger-mass disks, that planets will be produced. The suggested proportion of stars with planets, given by Steve as 0.34, does not seem excessive as an expectation from the nebula-theory model.

Solomon: The situation as I see it is that, while Steve has been able to produce estimates based on certain sets of parameters, since little is known about the statistics of the occurrence of these parameters it is not possible to give an overall estimate of the proportion of stars with planets, although the value estimated from observations is in the range given by some reasonable sets of parameters. Simon's case is different. He is pointing out that the existence of disks is indisputable, that their material content is more than enough to produce planets and that nebula-theory processes would inevitably produce planets from the material in a high proportion of disks.

Steven: If we are going to use arguments based on inevitability then the Capture Theory is based on the existence of dense embedded clusters, which are observational facts and which Simon accepts, on the analysis showing that capture-theory events are common and the simulations of the capture-theory process. Which of these elements does Simon find implausible?

Solomon: It seems to me that we can go no further on this topic. Where do we go from here I wonder? What do you two suggest?

Simon: I think that Steve and I should take turns in suggesting topics for discussion in the following weeks.

Steven: I'll go along with that.

Solomon: A good idea; I certainly would not know what to suggest. Since Simon made the proposal I think that Steve should choose the first topic. What will you choose, Steve.

Steven: One of the discoveries of the past few years is that the direction of the spin axes of stars may depart appreciably from being normal to the orbital planes of the accompanying planets — a phenomenon called *spin-orbit misalignment*. Comparing the ways that the Nebula Theory and the Capture Theory explain this should be interesting. It is most challenging to the Nebula Theory so I suggest that we use the following format. First, I will explain how the observations are made, then Simon can describe the way the observations are explained by the Nebula Theory and then, finally, I can give the capture-theory explanation. Will you agree with that Simon?

Simon: That's seems fine.

Solomon: This topic sounds as though it might be interesting — I look forward to next week. We have time on our hands. What have you two been up to this week?

The three friends chat until they leave and go their separate ways.

The Sixth Meeting

Steven and Simon Discuss Spin-Orbit Misalignment

It is 4th March 2016 and Solomon and Simon are sitting in the reserved cubicle. Steven approaches with three full glasses. He tells the other two: The landlord has recommended this — a new brew called 'Grant's Special'. He sits beside Simon and pulls a wad of papers out of a briefcase he had left on the seat.

Solomon: Last week we decided that you, Steve, are going to tell us about spin-orbit misalignment and how it is determined, then Simon will tell us how the observations are explained in terms of the Nebula Theory and, finally, you will give an explanation in terms of the Capture Theory. So, off you go, Steve.

Steven: It is self-evident that, on the basis of the Nebula Theory, the expectation would be, and was before observations to the contrary were made, that the spin axis of the star should be close to the axis of the orbital motion of the planets. The disk would form in the equatorial plane of the star and the planets form in the disk and orbit in its plane or, allowing for perturbation from some source or other, close to its plane. This condition is not exactly satisfied for the Solar System since the angle between the spin axis of the Sun and the normal to the average plane of the planetary orbits, what we call the *spin-orbit misalignment*, is about $7°$, not very large but also not negligibly small. Taken alone, this value for the Solar System does not pose any problem for the Nebula Theory; the probability of an alignment within 7° just by chance is 0.004, quite small, and one could think of various mechanisms — e.g. displacement of the planetary orbits by a nearby passing star — that could explain the tilt of the planetary orbits. We are going to use the term spin-orbit misalignment quite a lot this evening so I suggest we abbreviate it to 'SOM'.

It might be thought that the value of SOM would be impossible to estimate from observation of an exoplanet — it is difficult enough just to detect planets — but this turns out not to be so when the exoplanet transits its star. We can determine when a transit occurs because while the planet is passing in front of the star it blocks out some of the light it emits; from the proportion of light lost during the transit we can find the ratio of the radius of the planet to that of the star. The radius of a main sequence star is related to its temperature, which can be determined. The light emitted by a star is very close to black-body radiation so the wavelength of the peak of the emission curve is indicative of its temperature. Actually, astronomers have more subtle ways of estimating temperature, by measuring the relative intensities of a number of absorption lines in a star's spectrum, but we don't need to consider this further in our present context. All we need to know is that since we can find the radius of a main-sequence star we can also estimate the radius of the transiting planet. This diagram (Figure 6.1) shows the reduction in intensity as a planet transits the star HD 209458.

Most stars have a component of motion along the line-of-sight from the Earth so that the light coming from the star will be Doppler red-shifted if it is moving away from Earth and blue-shifted if it is moving towards the Earth; in principle the shift in the wavelength of the peak of the displaced emission curve indicates the star's radial

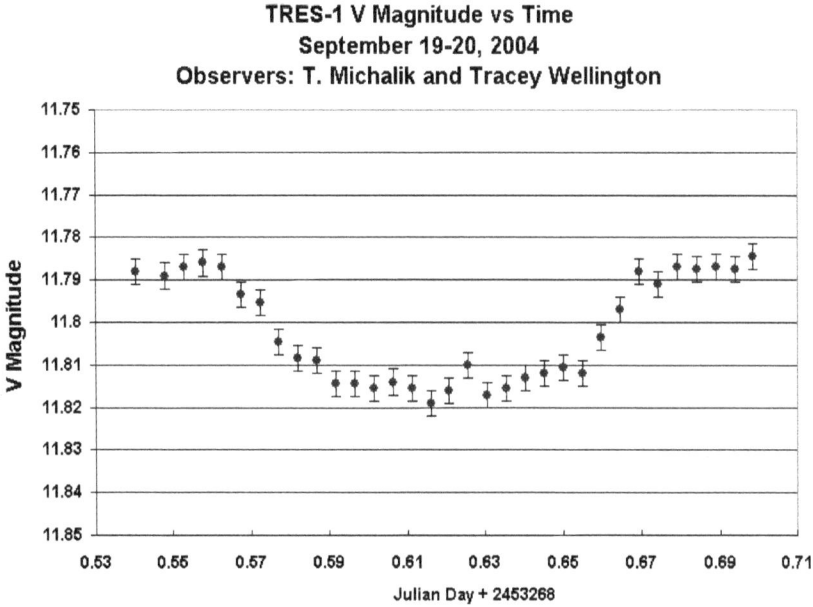

Figure 6.1   A typical transit recording

velocity with respect to the observer although, in practice, it is almost impossible to measure the tiny shift in a total spectrum. The red-or-blue shift is actually measured for spectral lines, which are well defined. In addition, because stars have an axial spin then, in general, some parts of the star move away from Earth relative to the star centre, and are thus more red-shifted, while other parts move towards the Earth relative to the star centre, and are thus more blue-shifted. We cannot see the disk of the star, since it is effectively a point source of light, but the effect of the spin is a slight spread of the emission curve relative to what would be coming from a non-spinning star. In this figure (Figure 6.2), the star is shown spinning about an axis perpendicular to the line of sight and the red and blue shading indicates red or blue-shift relative to the star centre. The black disk corresponds to the transiting planet in the equatorial plane of the star. The circle with a dot

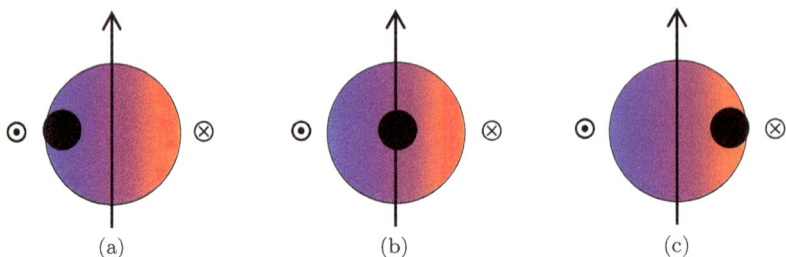

Figure 6.2   A spinning star and transiting planet

represents the head of an arrow, indicating that this side of
the star is moving out of the page and the cross indicates
the flight end of the arrow so that side of the star is moving
into the page; this indicates the sense of the spin of the
star. Going from left to right, the planet is orbiting the
star in the same sense as its spin.

At position (a), at the beginning of the transit, the
planet is blocking out the relatively blue-shifted part of
the light coming from the star and this is going to shift
the peak of the combined emission, or the mean position of
an absorption spectral line, towards red. Translating this
shift of wavelength into a radial speed would indicate that
the star was receding relative to the deduced motion of
the star when no transit was taking place. At (b), there
would be no apparent shift in radial speed and at (c), with
part of the relatively red-shifted part of the light removed
the peak of the combined emission would move towards
blue and indicate relative motion towards the observer
compared to the deduced motion in the absence of a
transit. Translated into apparent radial speed, relative to
the non-transit estimated speed one would obtain the time
sequence of relative speed as seen in (a) of this figure
(Figure 6.3), which indicates a SOM of 0°. The planet is
orbiting in the equatorial plane in the same sense as the
star. However, suppose that in this figure (Figure 6.2), the
planet orbits in the opposite sense, i.e. in order (c), (b),

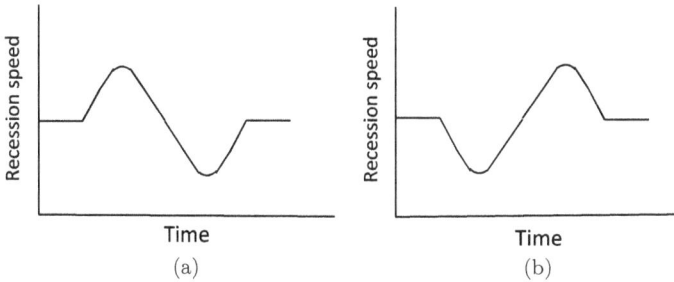

Figure 6.3 Estimated recession speed for (a) SOM = 0°; (b) SOM = 180°

(a), then the sequence of relative speed would be seen as in (b) (Figure 6.3), indicating an SOM of 180°, a retrograde orbit. This is the basis of the Rossiter–McLaughlin — shortened to RM — effect, first proposed in 1924.[1,2]

The RM effects illustrated in the two figures (Figures 6.2 and 6.3) correspond to very simple cases. However, in general the planetary orbit is not in the equatorial plane of the star and the spin axis of the star is not perpendicular to the line of sight. Since we can identify all the characteristics of a main sequence star from its temperature then we will know its diameter and from the observed orbital period of the planet we can deduce how long a transit across a diameter should take. Here I show some examples where the transit time will be shorter and from the time of transit we can find the angle between the diametric plane containing the orbit and the equatorial plane of the star (Figure 6.4). The figure shows the light curves for the three different kinds of transit.

Now, I don't want to get tied up with all the complex geometry involved in RM observations but, suffice to say, from the observations in favourable cases, it is possible to get an estimate of the SOM from observations of the

[1] Rossiter, R.A. (1924) *Astrophysical Journal*, **60**, 15–21.
[2] McLaughlin, D.R. (1924) *Astrophysical Journal*, **60**, 22–31.

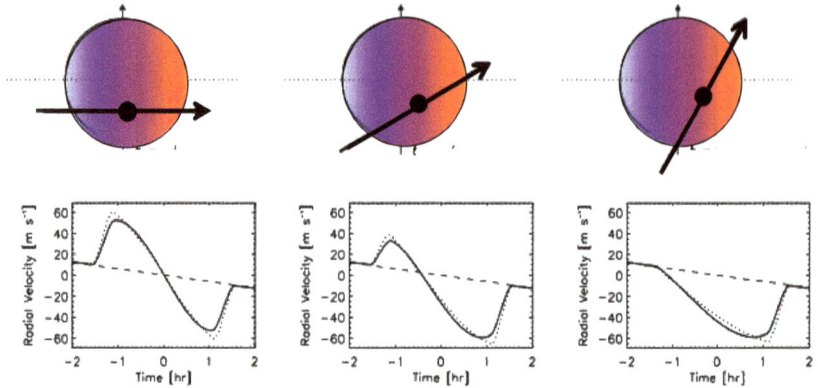

Figure 6.4   Three different kinds of transit and the corresponding RM curves

RM effect. The next figure (Figure 6.5) shows a histogram
of estimated SOMs for 71 transiting planets, including
Venus.[3] From this it can be seen that, while the complete
range of SOMs is possible, from 0° to 180°, there is a strong
bias towards low values. This is the observational data
that Simon must explain in terms of the Nebula Theory
and that I must explain in the context of the Capture
Theory.

Solomon: This looks like a very formidable challenge for any theory
but especially for the Nebula theory. I certainly had a
mental picture that all planets had to be close to the
equatorial planes of stars in direct orbits. I'll be interested
to see what answers the Nebula Theory can provide — so
off you go Simon.

Simon: Before we start, my glass is nearly empty and I am going to
need some lubrication as I explain how the Nebula Theory
deals with these observations. Drink up, Steve, and I'll get
some refills. I'll get the Grant's Special again; it's quite
good.

---

[3]Heller, R. (2013) http//www.aip.de/People/rheller/content/main_spinorbit.html.

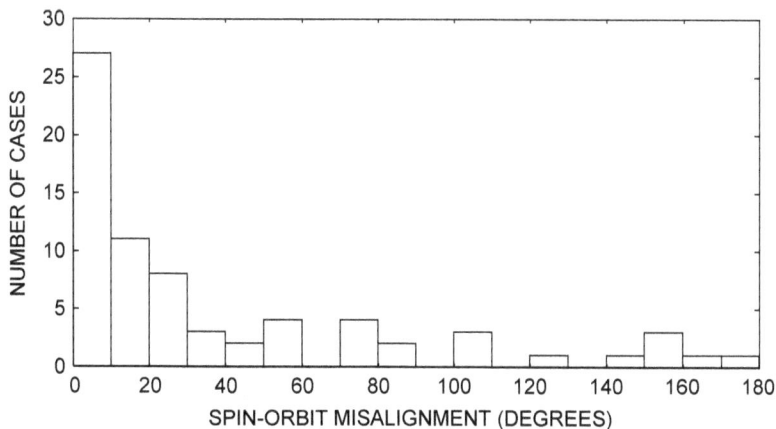

Figure 6.5   A histogram showing the SOMs for 71 planets in the Holt–Rossiter–McLaughlin Encyclopaedia

Steven drains his glass and Simon goes to the bar, returning after a few minutes with recharged glasses.

Simon: Now I can get started. In 2009, it was announced that the transiting exoplanet WASP-17b was in a retrograde orbit around its star and it was suggested that this was inconsistent with what would be expected from the Nebula Theory. This expectation that all SOMs should be zero, or very close to zero, is based on the idea that the development of the nebula takes place in isolation with the planets that are produced never interacting with each other or anything else. Well, we have already twice been told about the dense embedded state in the development of a stellar cluster, which means that some nebulae will evolve in the vicinity of compact stars, and also about migration processes that can propel planets outwards towards external stars. The idea of an isolated nebula and non-interacting planets is clearly not going to be the rule. Where it does occur then the spin and orbit axes will be aligned but there are a number of mechanisms that could give non-zero SOMs to a greater or lesser extent and I will now explain what they are.

The RM results were not exactly a shock to the solar-nebula community because the possibility of high-inclination exoplanet orbits was predicted in 2007 by Fabrycky and Tremaine[4] who considered the effect of the Kozai mechanism,[5] which involves perturbation of a satellite by a massive external body, which has the effect of giving an interplay between the orbital eccentricity and the inclination of the satellite's orbit. The maximum inclination achievable by this mechanism is about 40° and it can explain moderate SOMs through the influence of a star close to the evolving nebula. Another effect of an external star is the possibility that the circumstellar disk, in which planets form, could either be changed in orientation or be warped by stellar perturbation.

Another possibility involves the gravitational interaction of pairs of planets. Inward migration for planets of different masses will normally be at different rates, which can give rise to planets occupying close orbits and strongly interacting. There are many possible outcomes from such interactions. If the interaction is not too strong it can just perturb both planets into orbits of small-to-moderate inclination. To take another example, which I illustrate in this figure (Figure 6.6), in (a) we see planets A and B on orbits with slightly different speeds that bring them close together. In (b), we see the speeds with respect to a stationary centre of mass of the two planets. In (c), the effect of the interaction with respect to the stationary centre of mass is shown. With respect to the centre-of-mass frame of reference, both planets change their direction of motion. When the speed of the centre of mass is added, the net result can be either a reduction of speed in the original direction of travel or a retrograde motion for B. Planet A is given extra speed in its original direction of motion and

---

[4] Fabrycky, D.C. and Tremaine, S. (2007) *Astrophysical Journal*, **669**, 1298–1315.
[5] Kozai, Y. (1962) *Astrophysical Journal*, **67**, 591.

when this is added to the speed of the centre of mass it could have the escape speed from the star and be lost. The result of this kind of interaction depends very much on the relative masses of the planets and their original orbits. If A is more massive than B it can give a slingshot effect to B so that its backward velocity can be much larger than the original orbital speed and so give a net retrograde orbit. On the other hand, if A is in a high eccentricity orbit and the interaction is near its apastron, where its orbital speed is least, then the additional velocity in the forward direction due to the interaction can cause it to escape. Of course, many other interaction scenarios are possible, giving a range of final SOMs.

A quite different explanation for SOM was suggested by Rogers, Lin and Lau in 2012.[6] These workers noted that large SOMs occur mainly with very hot stars. These have very active convective cores and strong radiation from their outer regions. At the interface between the core and outer regions gravity waves are generated, which are waves in the material that attempt to restore equilibrium between the two regions; an example in everyday life are surface waves on the sea, known as *wind waves* at the interface between the atmosphere and the sea. Two-dimensional simulations have indicated the possibility of angular-momentum transport that will cause the surface material of the star to spin round a different axis from the main bulk of the star. On this interpretation, SOM does not really occur. The spin of the great bulk of the star is perpendicular to the orbital plane of the planets. It remains to be seen if three-dimensional simulations confirm this result.

You can see that there are many possible mechanisms to explain the observed SOMs, some or all of which may

[6]Rogers, T.M., Lin, D.N.C. and Lau, H.H.B. (2012) *Astrophysical Journal Letters*, **754**, L6–L10.

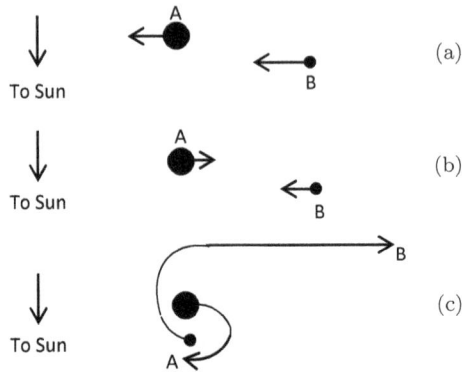

Figure 6.6   (a) Speeds of planets relative to star; (b) Speeds of planets relative to stationary centre of mass of planets; and (c) Speeds relative to stationary centre of mass after interaction

be operating. The observations that, while there is a bias towards small or zero misalignments, there are a large range of misalignments, including retrograde orbits, seem to be readily explained.

Solomon: You have given us plenty to think about — what do you think, Steve?

Steven: It is typical of many nebula-theory explanations for recent observations that they are very convoluted and neither easy to explain nor to understand. It is difficult to pinpoint specific problems with anything that Simon has given us on this topic — and given well I might add — but I just feel uncomfortable that so many mechanisms are invoked. However, that may just be the way it is.

Solomon: I thought that Simon made a convincing case. Is your story going to be any more straightforward?

Steven: I think so, but that will be for you to judge. Now, in the capture-theory scenario the star-protostar orbital plane could be in any direction and has no systematic relationship with the spin axis of the star. From the mechanics of the system, this will also be the orbital plane of the captured planets. On that basis alone, the initial values of SOM could be anywhere between 0° and 180°

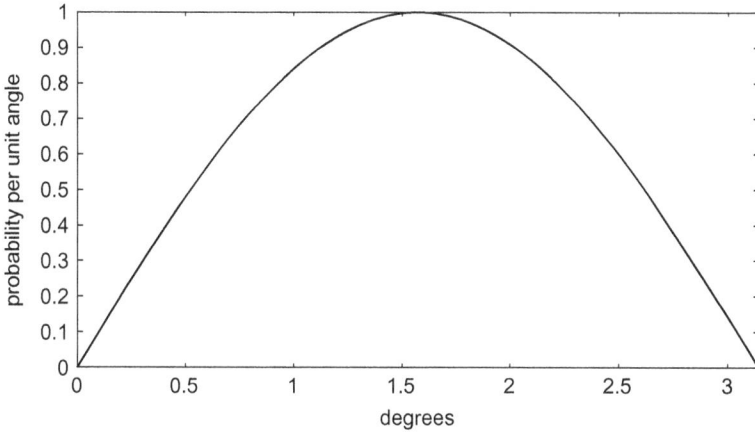

Figure 6.7   The relative probability per unit angle of initial SOM

although not with equal probability within that range. When you do the mathematics, the probability per unit angle looks like this (Figure 6.7).

I have emphasized that this gives the *initial* probability distribution because, as I shall show, there is another important process that comes into play. Three weeks ago I told you that some of the protoplanet material formed a disk around the star, typically of mass between 25 and 50 Jupiter units. What also happens is that some of that material impinges directly on the star, and this material imparts an angular momentum component appropriate to the rotation axis of the planetary orbits. Again, three weeks ago Simon told us about the Lynden-Bell and Pringle mechanism whereby if there is energy dissipation in a disk it will lead to inner material moving inwards and outer material moving outwards. In any gaseous disk, with gravity tending to produce Keplerian orbits and viscosity tending to equalize the velocities of neighbouring gaseous elements, there is bound to be some energy dissipation and hence some disk material drifting in to join the star. The addition of protostar material to the star, either directly from the protostar or via the disk, has the effect of shifting the star's spin axis from its original direction towards the orbital rotation axis and so to reduce the SOM.

It might be thought that it would take a large addition of protostar material substantially to affect the SOM but this is not so. The angular momentum of the present Sun is equivalent to about one-fifth of a Jupiter mass orbiting at its equator. Any disk matter spiralling in to join a star would, at the moment of being absorbed, be in free orbit around its equator but the angular momentum added per unit mass for material joining the star directly from the protostar would probably be greater. Thus, the addition of one or two Jupiter masses of protostar material could add angular momentum to the star with several, perhaps up to 10 or so, times the magnitude of the original angular momentum of the star. The effect of this can be seen here (Figure 6.8) for the case where the original SOM was 45° and the magnitude of the angular momentum of the added material is 5 times that originally in the star. The spin-orbit misalignment is reduced to 8.4°.

The range of initial magnitudes of stellar angular momenta and that of the captured material are unknown. However, just as an indication of the kind of overall results that are possible this histogram (Figure 6.9) gives the result of 1 million trials where the probability of the original SOM follows this distribution (Figure 6.7)

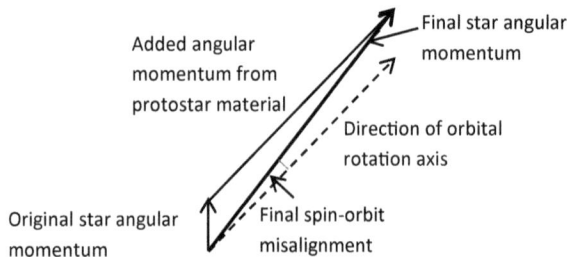

Figure 6.8   Due to captured material the SOM is reduced from 45° to 8.4°

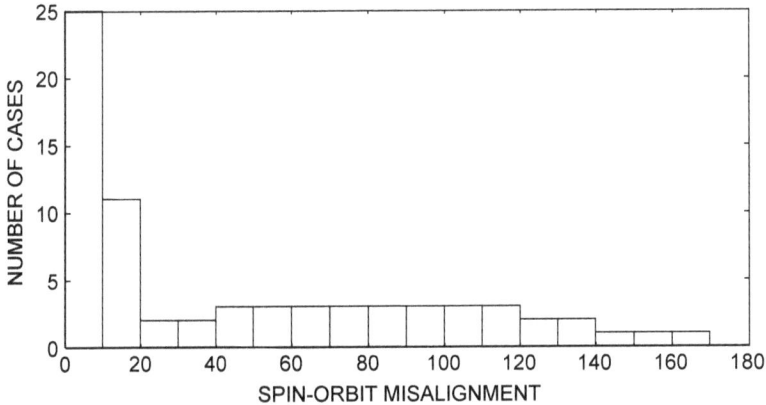

Figure 6.9   A histogram simulating 71 observations of spin-orbit misalignment where one-half of 1 million trials have the ratio of the magnitude of the angular momentum of captured protostar material to that of the star equal to 0.1 and the other half equal to 5.5

and where for half the trials the magnitude of angular momentum of added material is 5.5 times that of the star, while for the other half the ratio is 0.1. In each trial, the added angular momentum was added in a random direction. The numbers for each histogram block were scaled to give the equivalent of 71 observations and the result has a passing resemblance to the observed histogram (Figure 6.5).

There is an interesting difference between the nebula-theory and capture-theory explanations for the observed range of SOMs. The Nebula Theory begins with them being clustered tightly around zero and then finds mechanisms to spread then over the range 0–180°. The Capture Theory starts with a symmetrical distribution between 0 and 180° and describes a mechanism for giving a strong bias towards zero.

Solomon: Once again Steve, you seem to be able to present a more quantitative approach to explaining feature of the Solar System, but we must not read too much into that. As Simon has said, there is no requirement for astronomical processes that occur to always be amenable to simple analysis.

Simon: There is one observation of a SOM more than 170°. How likely is that in your scenario, Steve.

Steven: A good question. Let me think for a moment.

Solomon and Simon sit silently, sampling their ale, while Steven does a calculation.

Steven: I cannot answer your question with any precision since I don't have enough information. Any addition of protostar material to the star will reduce the SOM so the best-case scenario for an SOM greater than 170° is that there is no added protostar material. In that case, the probability that just by chance the SOM would be greater than 170° is easy to calculate; it is 0.0076. However, what is not known is the probability that the ratio of the angular momentum of added material to that of the star is zero or very small. If we guess that 10 percent of all CT interactions give very tiny additions of protostar material then the probability that our sample of SOMs would give 1 greater than 170° is (Steven writes)

$$P = 0.0076 \times 0.1 \times 71 = 0.054,$$

which is small but not negligible.

Solomon: This has been a good session. What occurs to me is that the range of possible scenarios in astronomy is so great that almost any observation can be explained — and often in more than one way.

Steven: I am very dubious about convoluted mechanisms that are counter to intuition. For example, can you really envisage a star where the surface material is spinning in a completely opposite sense to the material in the interior? I can tell you that, if observations had shown that all SOMs were very close to zero, I would have said that Capture Theory was wrong and given up working on it. I could, of course, have made the argument that so much protostar material is captured by the star that all SOMs are small but, since I wouldn't have believed such an argument, I wouldn't have made it. I really feel that the Nebula Theory is struggling with new observations. Each and every one is explained in some elaborate way — or sometimes in many elaborate ways since none is really convincing — but the cumulative effect of never having an easy explanation is that the theory is struggling.

Solomon: Well, that is your view; mine is more neutral. What do we discuss next week? Simon, it is your turn to choose.

Simon: Well, there is a problem with the Capture Theory that I would like Steve to explain — as far as I can see, producing planets in a filament can only produce major planets. Then, if we have any time after that we can talk about satellite formation. I can't quite see how it fits in with the Capture Theory.

Steven: That suits me. The formation of the terrestrial planets is something I wanted to talk about sometime.

They fill up the remaining time with Solomon describing a resent experiment at the Large Hadron Collider.

The Seventh Meeting

Steven Explains the Formation of Terrestrial Planets and the Earth–Moon Relationship by a Scenario Related to the Capture Theory and Then Simon Covers the Same Ground with the Nebula-Theory Interpretation

On 11th March 2016, the friends arrive together outside The Anchor, enter and occupy their reserved cubicle. Solomon goes to the bar and returns with three glasses of Grant's Special, the ale recommended by the landlord the previous week.

Solomon: If I remember correctly, Simon has challenged Steve to explain how terrestrial planets can be explained by the Capture Theory. I am not quite sure what the problem is — I assume that because they are closest to the Sun they were too hot to retain their atmospheres. Anyway, carry on Steve.

Steven: No Sol, that isn't the reason. There are major exoplanets much closer to their stars than Mercury is to the Sun, in fact about one-tenth of the distance, and they retain their atmospheres. What *is* true is that if you start with a planetary core of mass, say, a few Earth units very close to a Sun-like star then it would never be able to start to form an atmosphere; the escape speed of molecules from the core surface would be lower that the thermal speed of many of the molecules and they could not be retained. However, if a major planet formed further out and then migrated closer to the star it would be stable, since the escape speed from the whole planet would be much greater than that from the core alone. For example, the escape speed from the Earth is $11\,km\,s^{-1}$ but it is about $60\,km\,s^{-1}$ for Jupiter. The root-mean-square speed of hydrogen molecules at a temperature of 2,000 K is about $4\,km\,s^{-1}$ but, in the tail of the distribution, molecules will be moving much faster than the average so would escape from an Earth-mass core, or even a core somewhat larger and more massive.

Solomon: I see what you are getting at, and I can see that you do have to explain how terrestrial planets form. So tell us how.

Steven: I am going to start talking about something that seems to have nothing to do with terrestrial planets but, be patient, because what I'll tell you about not only relates to terrestrial planets but also to other features of the Solar System that I am sure we will discuss later. This topic is the ratio of deuterium to hydrogen in different kinds of body. When the Big Bang occurred, it mainly produced the light elements hydrogen and helium but also some of the hydrogen isotope, deuterium, with its extra neutron, and also traces of the next lightest element, lithium. The ratio of D/H in the Universe at large, in terms of relative numbers of atoms, is estimated to be about 2×10^{-5} but there are considerable variations within star-forming clouds and in bodies of the Solar System.

On Jupiter and Saturn the ratios are respectively 2.5×10^{-5} and 2.3×10^{-5}, slightly higher than, but not very different from, the accepted universal value. The values for Uranus, 5.5×10^{-5} and Neptune, 6.5×10^{-5} are both much higher. For the larger terrestrial planets the ratios are higher still, 1.5×10^{-4} for the Earth and for Venus the extraordinary high value of 1.6×10^{-2}. There is a special reason for the high Venus value. Venus originally had a large component of water vapour in its atmosphere but ultraviolet radiation from the Sun broke it up so that H_2O gave an OH radical plus H. The hydrogen would combine to give H_2 and two OH would give H_2O plus O, the latter of which would oxidize other things it came into contact with, like sulphur; there is a great deal of sulphur dioxide and sulphuric acid in the atmosphere of Venus. If the water molecule was HDO then the net product would be as just given previously except that instead of H_2 there would be HD. The lighter H_2 was more readily lost from the atmosphere than HD. This process did two things; it increased the D/H ratio and also led to a major loss of water so that Venus is now a relatively arid planet.

Recent measurements of D/H ratios in molecular species in either ice grains or ice-covered grains in star-forming clouds, and the protostars within them, have shown some very high values. In 2003, Roberts and his colleagues reported that in one cold dense cloud the ratio of doubly-deuterated ammonia, NHD_2, to normal ammonia, NH_3, is 0.05, implying that the D/H ratio in the ammonia in that cloud is 0.034.[1] In 2002, Parise and colleagues reported that in one protostar the D/H ratio in its ammonia is more than 0.02 and the amount of methanol, CH_3OH, either singly or doubly-deuterated, i.e. CH_2DOH or CHD_2OH, actually exceeds that of normal methanol.[2] Many other even higher D/H ratios in the molecules of icy grains in star-forming clouds have been found.

The reason for this concentration of deuterium in icy grains is a phenomenon known as *grain-surface chemistry*. A deuterium atom in the gas, falling on an icy grain, will dwell on the surface for a while. It may then swap places with a hydrogen atom in a molecule contained in the grain since changing hydrogen to deuterium in the molecule lowers its energy and physical systems are more stable

[1] Roberts, H., Herbst, E. and Millar, T.J. (2003) *Astrophysical Journal*, **591**, L41–L44.
[2] Parise, B., Ceccarelli, C. *et al.* (2002) *Astronomy and Astrophysics*, **393**, L49–L53.

with lower energy. Over the course of time the deuterium in the icy materials steadily increases, while that in the gas correspondingly reduces.

As a protostar collapses, slowly at first in free-fall, the dense grains sink to the centre, with the stone and iron components melting and segregating by density to form a core and mantle. The icy coatings of these grains are also carried inwards and, together with pure ice grains form a layer around the core-plus-mantle of deuterium-rich molecular material, which will be still present in the resultant YSO and beyond into the early main sequence state. I know that seems to be irrelevant to the formation of terrestrial planets but you will soon see that it is not.

Last week, in discussing SOM, Simon described how two planets, migrating inwards at different rates could interact and reverse the orbit of one to give a high SOM value and give enough energy to the other to enable it to escape. This requires a close interaction but something else that could take place is an actual collision between the two bodies. If this seems rather fanciful — although it shouldn't be — I will refer you to a NASA Spitzer Space Telescope report in 2009 of the observation of the debris of a planetary collision, which took place within the last few thousand years in the vicinity of the 12 million-year-old star HD 172555.

Now I will ask you to imagine that the early Solar System contained six major planets — the four that now exist plus two others that we will name as Bellona, the Roman goddess of war of 2.5 Jupiter mass, and Enyo, the Greek counterpart goddess, of 1.9 Jupiter mass. These masses are well within the range found for exoplanets. Each of the planets has been modelled for an SPH simulation in four layers on the basis that segregation by density in the planet had not gone to completion. The core consisted of iron plus some silicate, the mantle consisted of silicate with some iron, the next layer was deuterium-rich molecular material with some silicate and, finally, there was an extensive hydrogen-plus-helium atmosphere.

Three weeks ago, when I described the evolution of planetary orbits I told you that the final semi-major axis was inversely proportional to the mass of the planet. Another result of the calculation was to show that the larger the mass of the planet the faster its orbit decayed. With our large assumed masses for Bellona and Enyo their orbits would be the fastest to decay and hence they would have been the earliest to migrate into the confined region of the inner Solar System. I am going to describe a simulation that has been carried out of a collision between these two massive planets. Another important factor helping a collision to take place is that due to the mass of the gas-and-dust disk the gravitational forces on the planets have a non-central component, i.e. they do not point towards the Sun. This non-central force gives precession of the orbits, where, for slightly inclined orbits, the semi-major axis slowly rotates about the normal to the mean plane of the disk. The result of differential precession of two orbits, which may intersect in projection but not in three-dimensional space, is that the orbits do actually intersect from time-to-time thus giving the possibility of a collision. In 1977, Dormand and Woolfson, considering a set of six original solar-system planets, showed that the probability that a pair of planets would collide is of order 0.1, small but not negligible.[3]

This figure (Figure 7.1) shows the progress of the simulated collision and I'll take you through the various stages. The central dark regions are the silicate-iron core-plus-mantle. In (a), the planets are close to colliding and in (b) they have collided and merging of the atmospheres is occurring. In (c), some material, virtually all gas, is being

---

[3]Dormand, J.R. and Woolfson, M.M. (1977) *Monthly Notices of the Royal Astronomical Society*, **180**, 243–279.

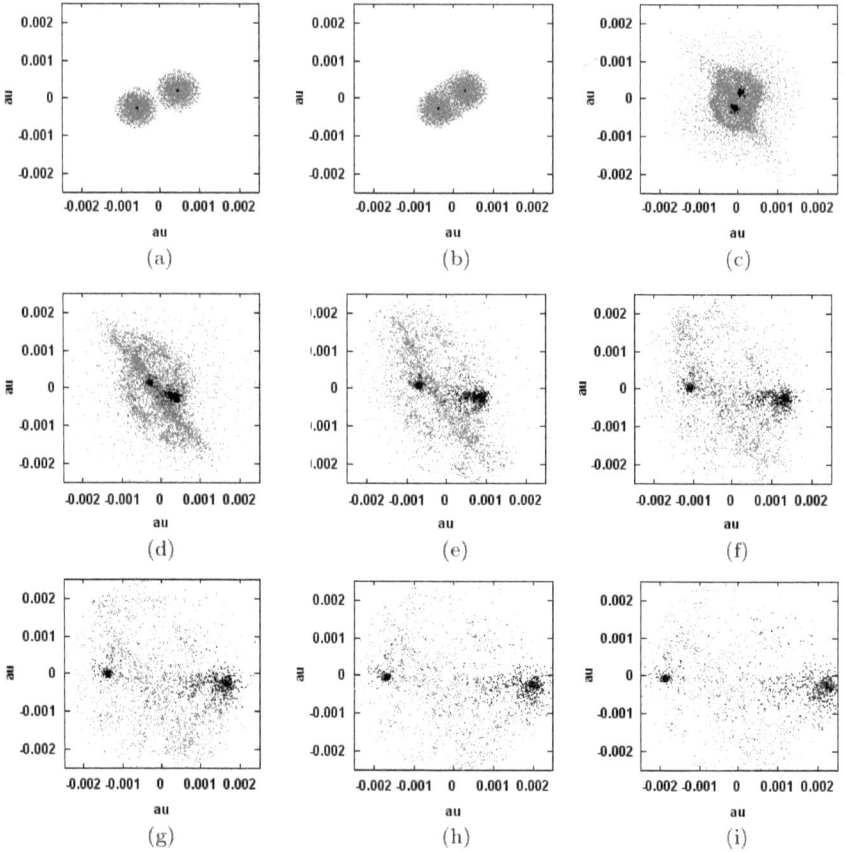

Figure 7.1   The progress of the collision. (a) $t = 0$, (b) $t = 590$ s, (c) $t = 1,326$ s, (d) $t = 2,505$ s, (e) $t = 3,917$ s, (f) $t = 5,336$ s, (g) $t = 6,415$ s, (h) $t = 7,597$ s, (i) $t = 8,609$ s (Woolfson, 2013)

ejected in the plane of the collision interface. Between (c) and (d), an important event occurs. Although most nuclear reactions require temperatures of about 100 million kelvin or more to take place, reactions involving two deuterium atoms occur at a high rate at a temperature of a few million degrees, the rate being also dependent on the density of deuterium. At this stage in the simulation the collision interface has reached the deuterium-rich region of Enyo and the temperature is just over 3 million degrees. Under these conditions D–D reactions occur in an explosive fashion and in (d) there can be seen the ejection of material from an explosion centre between the two cores. The following frames show the debris from the explosion spreading outwards, but also notice that parts of the two cores have survived and are moving apart.

These two cores are identified as proto-Earth and proto-Venus; with the collision close to the Sun their orbits can evolve to end up within the terrestrial region. In this particular simulation the mass of the Bellona core, identified as the proto-Earth is 2.5 times the mass of the Earth and that of the Enyo core, the proto-Venus, is about 1.5 times Earth mass. While these are larger than the actual masses of the Earth and Venus, the main conclusion from the simulation is that terrestrial-type planets can be formed in orbits within the terrestrial region of the Solar System.

It will not have escaped your notice that there are four terrestrial planets and that, so far, just the formation of the largest two has been explained. Let us look at the properties of these two small planets. Here I have prepared a table that shows their characteristics compared with those of the smaller of the larger terrestrial planets, Venus, and the largest and most massive satellite in the Solar System, Ganymede, one of the Galilean satellites of Jupiter (Table 7.1).

Given that both Bellona and Enyo were much more massive than Jupiter, and hence might be expected to have had more massive satellites, the proposal made here

Table 7.1   Characteristics of four smaller bodies

	Venus	Mars	Mercury	Ganymede
Mass (Moon units)	66.2	8.7	4.5	2.0
Density kg m^{-3}	5,243	3,933	5,427	1,936
Radius km	6,052	3,390	2,240	2,634

is that Mars and Mercury were two satellites released into heliocentric orbits that have partially rounded off to where they are now. I say partially rounded off because they have the largest eccentricities of any of the solar-system planets, 0.206 for Mercury and 0.093 for Mars, which might be expected for low-mass bodies because, as has been indicated previously, the larger the mass of a body the faster does its orbit evolve.

Now I have introduced the satellites that the colliding planets would have had, I would like to discuss them further. There would have been satellites of large mass, such as those that became Mars and Mercury, but many smaller ones as well; both Jupiter and Saturn have many tens of satellites, although some may be bodies captured long after the planets formed. To start with there is Ceres, with radius 473 km and orbit with eccentricity 0.076, situated in the gap between Mars and Jupiter. Since its discovery in 1801 it was first designated as the largest asteroid but now is labelled as a *dwarf planet*, a class of objects in heliocentric orbits, large enough for self-gravitational forces to give them a near-spherical shape but not with sufficient mass to clear the neighbourhood of its orbit. Now it is suggested that Ceres is an ex-satellite of one of the colliding planets.

Another possible outcome for a satellite after the collision is that it is retained as a satellite of one or other of the surviving cores, which have been identified as becoming the larger terrestrial planets. It is now proposed that the Moon is a satellite retained by the Bellona core — the Earth. Is there any evidence that this might be so? Actually there is. In 1959, the Soviet Union spacecraft, Lunik 3, photographed the far side of the Moon and revealed the surprising fact that it is unlike the side that faces the Earth. The side we see is covered with large mare basins, more-or-less circular regions representing enormous craters that have been produced by large projectiles and have subsequently been partially filled by magma welling up through cracks in the floor of the basins from the interior of the Moon. Between the Mare are highly cratered highland regions. The far side is quite different being almost completely highland regions with one significant, although comparatively small, mare basin, Mare Moscoviense. This photo shows parts of the two sides and the boundary marking the hemispherical asymmetry of the Moon is clearly seen (Figure 7.2). Of course, planetary scientists were soon trying to explain this phenomenon and the first thought was that there had been asymmetric bombardment of the Moon — perhaps the Earth had focussed projectiles on the near face of the Moon so that is was preferentially bombarded. Computer simulations did not support that theory but, in any case, altimeter readings from spacecraft orbiting the Moon showed that there were large impact basins on the far side of the Moon — they just hadn't filled up with magma from below. The reason given for this was that during the time of heavy bombardment by large projectiles the crust of the far side had been thicker so that the impact

Figure 7.2   Parts of the two different hemispheres of the Moon

basins, or rather the cracks in the basin floors, did not penetrate into the region of molten magma. Measurements from seismometers left on the Moon's surface by various American Apollo missions have confirmed this conclusion; although the crust over the Moon's surface is of variable thickness the crust on the nearside is, on average, a few tens of kilometres thinner than on the far side.

That is the answer to the problem of explaining the hemispherical asymmetry of the Moon, but this is an answer that raises another question. Why was the solid crust thinner on the nearside? Now, we know that, because of tidal effects, satellites have orbits around their planets such that the orbital period equals the spin period, so, that one face always points towards the planet — so-called synchronous orbits. So, it would have been for the Moon and Bellona. However, an early satellite, at a high temperature with a fluid interior and somewhat plastic outer regions, should have had a *thicker* crust on the nearside (Figure 7.3). Due to tidal forces, the low density crust would pile up in the direction of the planet much as a sea tide piles up on the Earth towards the Moon. On the Earth, there is also a tide on the side away from the Moon but for a satellite very close to a planet that effect would be smaller.

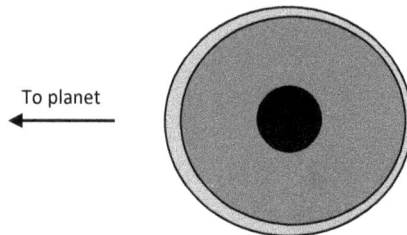

To planet

Figure 7.3   The initial structure of a satellite formed in synchronous orbit around a planet showing core (black), mantle (dark grey) and crust (light grey). The thickness of the crust is exaggerated

The debris from the collision, clearly seen here (Figure 7.1) is travelling at more than $100\,\mathrm{km\,s^{-1}}$ and it can be shown that the amount falling on the hemisphere facing the collision could abrade a few tens of kilometres from the surface,[4] turning what was the side with a thicker crust into the one with a thinner crust. When a period of heavy bombardment of the Solar System occurred, some 500 million years after its formation, only the nearside of the Moon could give mare basins, such as we see. The internal structure of the Moon that resulted from its relationship with Bellona would also have ensured that the bombarded side faced the Earth. I'll talk later about the period of bombardment and what might have caused it.

Solomon: I find that very interesting and it is all quite new to me. I have never seen this idea in any of the popular articles I have read about the Solar System. Carry on Steve, I wonder what new ideas you still have to offer.

Steven: No, all you see in popular literature is nebula-theory ideas; the juggernaut rolls on despite the fact that the wheels are falling off! Before continuing I would like a short break while I cure my thirst and get some refills.

They empty their glasses and Steven goes to the bar, coming back after a few minutes with a tray of full glasses.

Solomon: Carry on Steve — I'm agog to hear what's coming next.

Steven: Don't get too excited. All this is speculation and it can only be assessed on the basis of plausibility. Nothing can be considered as 'right' in this field although we can say something is wrong if it is clearly in conflict with observation. Anyway, to continue. Another body showing clear hemispherical asymmetry is Mars as we can see here (Figure 7.4). In this case, instead of a number of maria on

---

[4]Woolfson, M.M. (2013) *Earth, Moon and Planets*, **111**, 1–14.

Figure 7.4   The topography of Mars showing the northern plains (blue end of spectrum) and the southern highlands (red end of spectrum). The deep blue region in the south is the Hellas Basin, caused by a large and energetic projectile

one side we have abrasion to such an extent that a whole hemisphere has been flooded with magma. This may be due to a more intense bombardment or to the fact that Mars, being a much larger body than the Moon, would have had molten material closer to the surface.

There is another important difference between the arrangement of the hemispherical asymmetry in the Moon and Mars. The Moon was retained by the Bellona fragment — the Earth — and the dynamics of the situation ensured that the plane of asymmetry contains the spin axis in Earth orbit as it did in Bellona orbit. However, in the case of Mars, the plane of asymmetry makes an angle of 56° with the spin axis. Mars became a body that was essentially isolated from any other body. Because of internal motions of viscous fluid material its energy would have decreased while, because it was isolated its spin angular momentum would remain constant. Solid surface material was able to slide over the molten mantle, somewhat like continental drift on Earth, and, to satisfy the energy and angular momentum requirements, the arrangement of material had to satisfy a physical condition that has a formal scientific formulation but can be described verbally as material migrating as far as possible from the spin axis.[5] Modelling the detailed topography of Mars, as a set of elevated and depressed regions relative to the mean level of the highland region, this condition has been shown to be closely satisfied.[6]

When we look at this table (Table 7.1), we can see that Mars fits between Venus and Ganymede in terms of both mass and density but that Mercury does not do so for density. If one allows for the compression of a planet due to the pressure provided by its own mass then Mercury is intrinsically denser than any other planet, including Earth that, in simple terms, means that it has the highest proportion of iron. Models of Mars and Mercury show that they have similar size iron cores so that they largely differ only in the amount of silicate they have in their mantles. For this reason, it has often been speculated that Mercury was once a Mars-size body that had a great deal of its mantle stripped away in a collision with another body. If Mercury were close to the colliding planets, especially within the extended plane of the collision interface where debris would be particularly dense, then much of the mantle facing the collision could have been stripped away. There would have been a re-assembly of the remaining material to form a spherical body in hydrostatic equilibrium. It may be significant that there are two regions of Mercury in antipodal arrangement, i.e. they are on opposite sides of a diameter, which have special characteristics. One is the Caloris Basin, which looks like a large impact feature, where there are circular ripples in the surface like those

[5] Lamy, P.L. and Burns, J.A. (1972) *American Journal of Physics*, **40**, 441–448.
[6] Connell, A.J. and Woolfson, M.M. (1983) *Monthly Notices of the Royal Astronomical Society*, **204**, 1221–1230.

in a pond when you throw in a stone, and at the other end of the diameter there is a rough region referred to as Chaotic Terrain. These regions face the Sun at alternate perihelia — the ratio of the period of Mercury's orbit to its spin period is 3:2. Here is a possible sequence of events that could have given these surface features (Figure 7.5). Frame (a) shows the arrival of high-speed debris and (b) the stripped mantle. In (c), the mantle flows round the remaining body and overshoots to form a plume. The flow back shown in (d) might give rise to something like the Caloris Basin and the Chaotic Terrain could be due to the meeting of this material when it reached the far side producing a smaller overshoot, as we see in frame (e). All this is very speculative and needs more investigation to

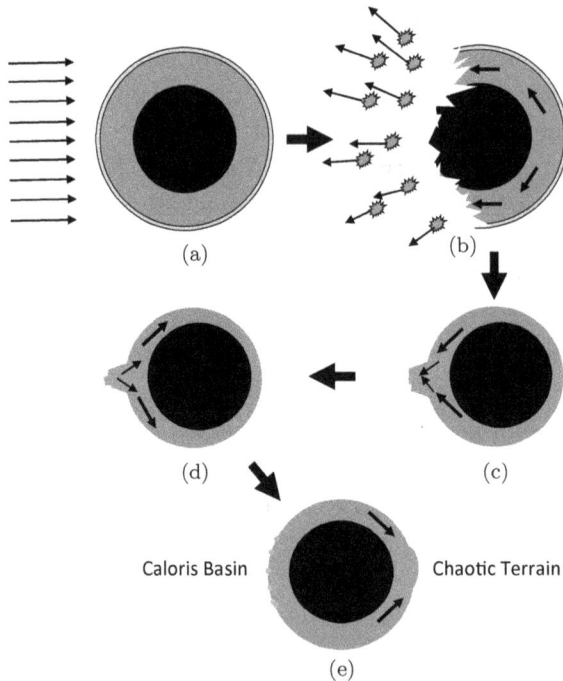

Figure 7.5   Stages in the development of Mercury

see if it is feasible, but it does give a possible explanation for what is observed.

Solomon: You have covered a lot of ground there Steve and I found it interesting. However, although you have explained the nature of the terrestrial planets and features of the Moon and smaller terrestrial planets I still have a feeling that it is all a bit *ad hoc* — that is you have invoked a planetary collision out of the air, so to speak, just to get over a difficulty of the Capture Theory.

Steven: I don't think that is quite fair Sol. Remember, there is observational evidence of a planetary collision somewhere else in the galaxy; if they were very rare events it is unlikely that you would observe the results of one that took place just a few thousand years ago, a blink of an eye in astronomical timescales. Later I will show you that the collision explains much more, and a proposed mechanism that explains a large number of features of the Solar System can hardly be described as *ad hoc*.

Solomon: Alright, I'll suspend judgement for now. Well Simon, what is the nebula-theory take on terrestrial planets and the Moon.

Simon: I'll start with the Moon because there is a well-modelled scenario for how this came about and one that is generally accepted by the astronomical community. For now, we'll accept that the Earth exists as a rocky body and that the inner Solar System contains a number of bodies of diameter 1,000 km or more, formed by the accretion of planetesimals as I described last month. In 1987, Benz, Slattery and Cameron modelled what would happen if a Mars-size body struck the Earth a glancing blow.[7] This figure shows the result of their SPH simulation (Figure 7.6). Both the Earth and the impacting body are represented by an inner iron core and a silicate mantle.

---

[7]Benz, W., Slattery, W.L. and Cameron, A.G.W. (1987) *Icarus*, **71**, 515–535.

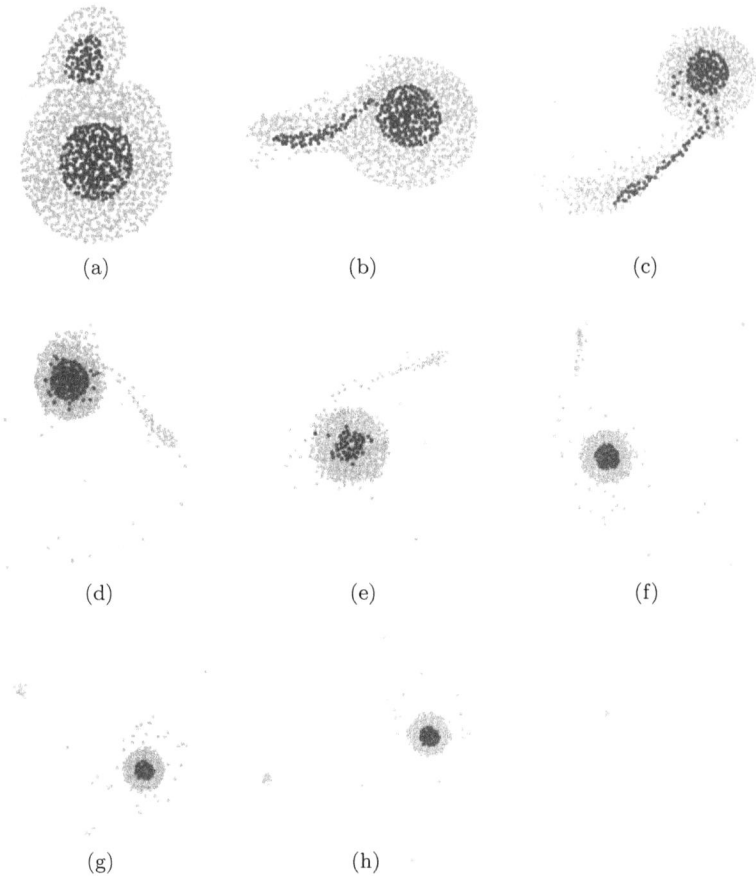

Figure 7.6   The impact model of Moon formation (Benz *et al.*, 1987)

After the initial impact, the impacting body is stretched into a filamentary form, leaving some of its material incorporated into the Earth. The material at the outer end of the filament collapses to form the Moon, as seen in frame (h). It is generally agreed that this gives a very satisfactory explanation for the formation of the Moon and its relationship to the Earth.

Steven: Can I interrupt at this stage?

Solomon: OK, but keep it short.

Steven: Firstly Sol, you criticized my planetary collision as being *ad hoc*, I'll show that you are wrong in that respect but if you want an example of *ad hoc* then this is it — a collision that is postulated for one purpose only — the explanation of the Moon's origin and relationship to the Earth. However, it gives no obvious explanation of the hemispherical asymmetry of the Moon and, actually, the Benz, Slattery and Cameron model gives a Moon with too little iron in its core.

Simon: The amount of iron can probably be adjusted by varying the parameters of the collision and, since the Moon condensation formed at the end of a filamentary structure there could have been an asymmetry in its formation leading to a variation of crust thickness. Anyway, let me now return to the formation of terrestrial planets.

Steve has already pointed out that inward migration in a resisting medium is faster for more massive bodies, In 1998, two Japanese astronomers, Kokubo and Ida pointed out that when a collection of planetesimals forming in the inner Solar System reached about the mass of the Earth, its inward migration would be so fast that it would very quickly move in and join the Sun.[8] Hence, to form the terrestrial planets, we have to wait until the gaseous nebula has either completely disappeared or is so tenuous that its effect on the orbits of bodies within it is minimal. At that stage, there will still be many planetesimals present and the terrestrial planets can form and not be sucked into the Sun. Venus and the Earth formed as they are now but Mars grew under the handicap that the nearby presence of Jupiter is stirring up the planetesimals in its vicinity and so inhibiting growth — hence its small mass, about one-ninth that of the Earth. As Steve has already told you, Mercury is thought to be the residue of a

---

[8]Kokubo, E. and Ida, S. (1998) *Icarus*, **131**, 171–178.

Mars-mass object that was struck by some other body and had much of its mantle stripped off. Such a collision would also have displaced it so it cannot be said with certainty where it originated, but its large orbital eccentricity could indicate that it was propelled into an elliptical orbit that did not round-off completely.

There is not a lot more I can say about terrestrial planet formation. It seems a very straightforward process once the resisting medium has either gone completely or been reduced in its effectiveness to modify planetary orbits.

Steven: Again, no explanation for a very important feature of Mars — its hemispherical asymmetry. Again, how did the Earth and Venus reach their near-circular orbits without the presence of a resisting medium? They could hardly have formed in those orbits.

Simon: There could be a number of scenarios envisaged where there was asymmetric bombardment of Mars. As far as the orbits of the Earth and Venus are concerned then dynamical friction due to interactions with residual planetesimals can give round-off in much the same way as the resistance of a gaseous medium.

Solomon: We've had a pretty full session this week and you have both put forward pretty good theories about the origin of the terrestrial planets and the Moon. Since the Moon has been featured I am quite interested to know how you explain the formation of satellites. There are some much larger than the Moon and many much smaller. How do they come about?

Steven: That will be a good topic for next week. Do you agree Simon and will you kick off the discussion?

Simon: Yes, indeed.

They drink up, chat for a while and then go their separate ways.

The Eighth Meeting

Simon and Steven Talk
about Satellite Formation

On 18th March 2016, Steven and Solomon are sitting in their reserved cubicle. Simon comes to the table with three glasses of Grant's Special and, after sampling the brew, Solomon begins talking.

Solomon: This week you are going to tell me about satellite formation. As I was travelling here I realized that satellites are related to planets as planets are to stars so maybe I am not going to hear very much that is new. We'll see if I'm right. Simon, off you go!

Simon: Before describing how satellites form, it is first necessary to say something about them because they have different characteristics and they are not all formed in the same way. There are two main kinds of satellite — regular and irregular. Regular satellites tend to be closer to the planet and they have orbits that are circular, or very nearly circular, and close to the equatorial planes of the planets. The only satellite we have considered so far, the Moon, is clearly not a regular satellite; its orbit is inclined at more than 5° to the Earth's equator and its orbital eccentricity is 0.054. Last week I described the impact hypothesis of the origin of the Moon and Steve's model made it a retained satellite after a planetary collision.

There are many other irregular satellites; some are small, others at great distances from planets and quite a

few are in retrograde orbits, e.g. they orbit in the opposite
sense to the spin of the planet. The irregular satellites may
have many different kinds of origin and I will mention some
of these. However, explaining the regular satellites is far
easier so I will deal with these first.

There are two main ideas about how regular satellites
could form. The first of these is that when the planetary
cores accrete gas to form the atmospheres of major
planets a disk also forms, within which regular satellites
are produced in much the same way that planets are
produced around stars, except that everything happens
on a much smaller scale. That model would support what
you said, Sol, about not hearing anything new. Another
idea, put forward by an international team of mainly-
French astronomers in 2011, is that the formation of
regular satellites is intimately connected to the presence
of rings around the major planets.[1] Everybody is aware
of the substantial ring system around Saturn but one of
the discoveries of the space age is that Jupiter, Uranus and
Neptune also have rings, although they are so tenuous they
can only be imaged directly by spacecraft passing close
by. This figure (Figure 8.1) shows the rings of Jupiter, the
top image, showing more structure, was taken with back-
scattered light and the lower image with forward-scattered
light.

Actually the rings of Uranus were detected in 1977,
well before they were imaged by the Voyager 2 spacecraft
in 1986. Three American astronomers, James L. Elliot,
Edward W. Dunham and Jessica Mink, planned to use the
Kuiper Airborne Observatory, a telescope facility operated
from an aircraft flying at a height of 14,600 m, to study
the atmosphere of Uranus by observing the occultation
of a star passing behind it. The star would dim as it

---

[1] Charnoz, S. *et al.* (2011) *Icarus*, **216**, 535–440.

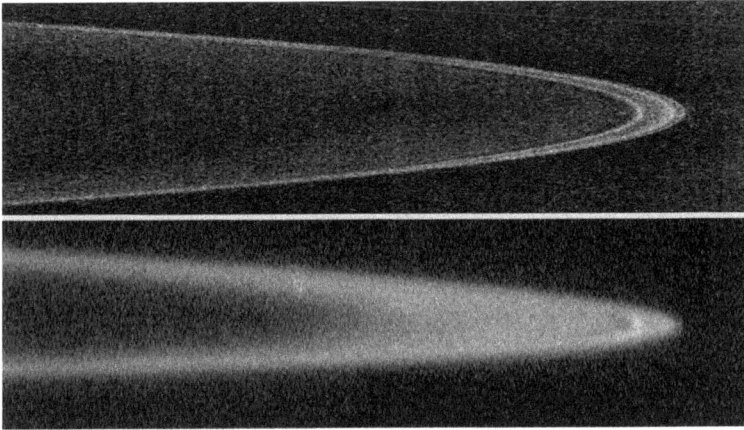

Figure 8.1   Two images of the rings of Jupiter (NASA)

passed behind the atmosphere and be obscured completely as it passed behind the solid planet. However, before the star reached the atmosphere it repeatedly disappeared and this was interpreted as due to its passage behind bodies contained in a ring system surrounding Uranus. Nine rings were detected in this way.

The rings of Saturn consist of small icy bodies and they cannot come together to form an icy satellite because such a satellite would be within the Roche limit. This is the distance at which a large body would be torn apart by tidal forces, caused by the differential pull on the two sides of the body due to the gravitational field of the planet, over-coming the self-gravitational forces holding the satellite together. Tidal forces depend on the size of the body being affected — the larger the body the greater the tensional forces across it — so small bodies can survive without being disrupted because their mechanical strength is sufficient to resist the tension due to the tidal force. The assumption is that the rings spread over time and as regions of the rings migrate beyond the Roche limit so satellites would be able to form by the gradual accumulation of bodies in the ring. According to this theory, for Jupiter, Uranus and

Neptune, the flimsy existing rings are the residues of much more substantial ring systems that were consumed by the satellite-formation process. It is claimed that the model accounts for the fact that, for the satellite systems of the major planets, the larger satellites are further from the planets; since they were the first to form they would have had longer to grow.

As far as irregular satellites are concerned it is impossible to give a single mechanism that would explain them all — they are so individual in their characteristics. There are two main types of mechanism that can give rise to the capture of a body by a planet, both of which require the removal of energy from the body so that it is slowed down to below the escape speed from the planet. The first is that there should be some other body present in the vicinity, preferably a substantial one, which could take up some of the energy of the body to be captured. The second is for a collision to take place: if two bodies collided head on in the vicinity of the planet then a considerable part of the kinetic energy of the bodies would be taken up in shattering the bodies and producing heat.

Just as an example we take Phoebe, the very curious irregular satellite of Saturn. It is a roughly-spherical, but highly-pockmarked, body of average radius 108 km, the sort of size one might expect for a large planetesimal. Planetesimals are expected to be of irregular shape since a body of radius less than about 300 km does not exert enough self-gravitational force to overcome the mechanical strength of its material and hence become spherical. However, if Phoebe formed very early in the formation of the Solar System then it would have contained some short-lived radioactive isotopes, in particular aluminium-26 that has a half-life of 717,000 years, giving magnesium-26 as the end product. The energy generated by the decay over such a short time would have made the interior of Phoebe molten and enabled it to take up a near-spherical shape.

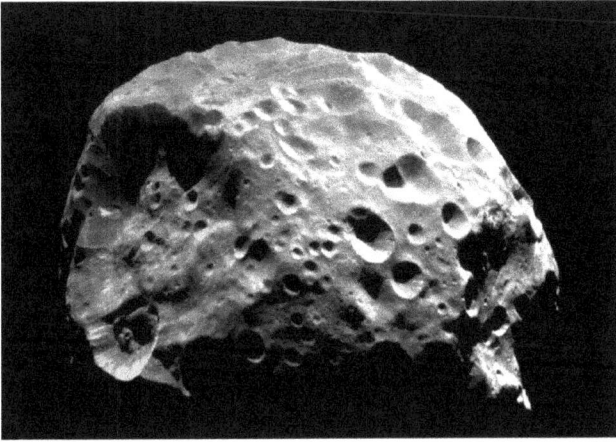

Figure 8.2   Saturn's irregular satellite, Phoebe (NASA)

Its orbit has a semi-major axis of 12.9 million km, nearly 4 times as large as that of Iapetus, the next satellite inwards, and an eccentricity 0.156, The inclination of its orbit relative to the equatorial plane of Saturn is 151°, meaning that it is in a retrograde orbit inclined at 29° to the planet's equatorial plane. Here is a spacecraft image of Phoebe (Figure 8.2). It is a very heavily bombarded body and seems to be covered with a dark deposit of some kind. There are two very large impact features, one on the left side and the other on the right side, both of which are indicative of a major collision. An interpretation of its visual appearance is that it was a planetesimal in a heliocentric orbit that was struck head-on by a smaller planetesimal in the vicinity of Saturn. This slowed it down relative to Saturn to the extent that it was captured. The outer regions of Phoebe are thought to be ice but the surface became covered by a layer of dark material consisting of dust from the colliding planetesimal.

Other scenarios can be constructed to explain other irregular satellites, although one cannot be sure that they are always correct. Anyway I think that I ought to hand over to Steve at this point.

Steven: I think that Simon and I are pretty much in agreement
about satellite formation. I am going to describe a semi-
quantitative study of how regular satellites form but before
I do that I want to follow up on Simon's description of the
formation of Phoebe by describing how some interesting
irregular satellites of Jupiter may have formed.

There are two groups of small satellites at considerable
distances from Jupiter. The members of the inner group,
orbiting with semi-major axes between 11 and 12 million
km, have very eccentric orbits with inclinations between
25 and 29°. The members of the outer group, orbiting
with semi-major axes between 20 and 24 million km, have
retrograde orbits with inclinations between 145 and 165°
and have even higher average eccentricity than the inner
group. They are the result of some capture process and it is
significant that the apojove[2] distance, 14.18 million km, of
Elara, a member of the inner group, is just greater than the
perijove distance, 13.75 million km, of Pasphaë, a member
of the outer group. These two groups were probably caused
by the collision of two asteroids, which Simon would refer
to as planetesimals, in the vicinity of Jupiter. A suggested
scenario is shown in this figure (Figure 8.3). The asteroids
A and B were both originally in direct orbits around
the Sun. After they collided, there were two groups of
fragments following paths that deviated from those of the
parent asteroids, although they were in the same general
direction. The inner group were, and are, in direct orbits
and the outer group in retrograde orbits.

However, what I mainly want to talk about is the
formation of regular satellites. The fact that all four major
planets in the Solar System have regular satellites makes
it likely that satellite formation is an almost inevitable
concomitant of major-planet formation. Will we ever be

---

[2]Perijove and apojove are the minimum and maximum distance from Jupiter for an orbit
around that planet.

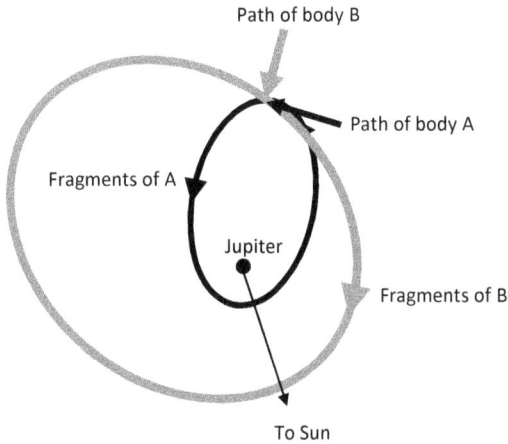

Figure 8.3 A collision of two bodies near Jupiter giving rise to two outer families of satellites. All motions are shown relative to Jupiter

able to detect satellites of exoplanets I wonder? Anyway, the prediction is that exoplanets would have satellites even it can never be confirmed.

Last month, when I described the formation of planets by the capture-theory process, I produced this computer-generated figure showing the collapse of a protoplanet (Figure 2.9). What you will notice is that the collapsing centre of the blob, which will go on to produce a planet, is surrounded by a disk and it is in this disk, which has a mass only a little less than that of the collapsing core, that satellites can form. Before giving details of this process I would first like to make some general comments concerning satellite formation.

When Galileo first saw the large satellites of Jupiter through his telescope in 1608 he interpreted it as a small-scale version of the Solar System and it reinforced his belief in the Copernican Sun-centred model. For him, and many others that followed, it became almost axiomatic that the mechanism for producing satellites should be a small-scale version of that for producing planets.

Table 8.1   The ratio, S, of the intrinsic angular momentum of the secondary orbit to that of the spin of the central body at its equator

Central body	Secondary body	Ratio S
Sun	Jupiter	7,800
Sun	Neptune	18,700
Jupiter	Io	8
Jupiter	Callisto	17
Saturn	Titan	11
Uranus	Oberon	21

We have had a lot to say about the importance of angular momentum and the difficulty it presents to nebula-type theories because the planets, with 0.0014 of the total mass of the Solar System, contain in their orbits 200 times as much angular momentum as the Sun does in its spin. This problem does not exist for the Capture Theory since the angular momentum of the Sun, or a general star, is a result of the star-forming process while the much larger angular momentum in the planetary orbits is derived from the Sun-protostar orbit. Here, in this table (Table 8.1), I compare the planetary system and satellite systems in terms of where angular momentum resides in them. It shows the following ratio with respect to a number of primary and secondary pairs of bodies. With 'intrinsic' meaning 'per unit mass' this ratio is (Steven writes on a sheet of paper)

$$S = \frac{\text{intrinsic orbital angular momentum of secondary body}}{\text{intrinsic angular momentum for material at primary equator}}.$$

In the table this quantity, S, is given for various pairs of bodies, which clearly shows a distinct difference between the planetary system and satellite systems. The intrinsic orbital angular momentum of satellite systems does not dominate to the same extent as it does for the planetary system. Because of this fact, it is not unreasonable to consider a different mode of formation for satellites around planets from that for planets around stars — indeed, it might be unreasonable to do otherwise.

Solomon: Sorry to interrupt, Steve. But I think that it is about time for refills. Finish your drinks.

They spend a few minutes drinking and chatting. When the glasses were all empty Solomon went to the bar and returned with refilled glasses a few minutes later.

Steven: Now, for a change, Simon and I are going to be in agreement because I am going to describe a process of satellite formation that exactly parallels the process of planet formation proposed for the Nebula Theory. The steps are:

(i) Dust settles into the mean plane of the protoplanet disk to form a dense dust carpet.

(ii) The dust carpet is gravitationally unstable and fragments to produce solid bodies — which we call satellitesimals.

(iii) The satellitesimals aggregate to form satellites.

I will start with the settling of dust into the mean plane of the disk. Although most dust particles in molecular clouds are of submicron size, which are the easiest to detect because of the way they scatter and absorb light, there is a distribution of sizes going up to about 5 microns. The numbers of dust particles, of diameter D, per unit increment of diameter falls of quite sharply, as $D^{-3.5}$, so there are 280 dust particles with diameter between 0.99 and 1.00 microns for every one with diameter between 4.99 and 5.00 microns. From this distribution, we find that more than one third of the mass of dust is contained in particles above 2 microns in diameter. The larger particles move most rapidly towards the mean plane of the disk and as

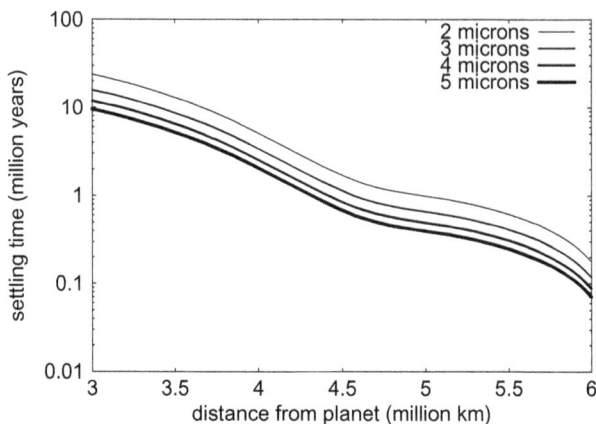

Figure 8.4   The settling time for dust at different distances from the planet and various particle diameters

they do so they sweep up smaller particles they encounter that are falling more slowly.

The disk will have a flared structure, as I showed previously and show here again (Figure 4.4), and an areal density decreasing approximately exponentially outwards from the planet. Particles closer in will have less far to fall but do so in a denser medium that offers more resistance to motion. Settling times at different distances from a planet are shown in this figure (Figure 8.4), using the theory developed by Weidenshilling, Donn and Meakin,[3] for the following set of parameters:

Mass of planet $= 2.0 \times 10^{27}$ kg (approximately the mass of Jupiter),
Mass of disk $= 2.0 \times 10^{27}$ kg,
Areal density fall by factor e every 2×10^6 km distance from planet,
Mean molecular mass of gaseous material $= 4 \times 10^{-27}$ kg,

---

[3] Wiedenschilling, S.J., Donn, B. and Meakin, P. (1989) *The Formation and Evolution of Planetary Ststems*, Eds. H.A. Weaver and I. Danley (Cambridge University Press, Cambridge).

Temperature of disk material $= 20\,\text{K}$,
Density of dust grains $= 3 \times 10^3 \text{ kg m}^{-3}$,
Ratio of principle specific heats of gas $= 5/3$.

At distances beyond 4.0×10^6 km, for a postulated disk lifetime of 3 million years, settlement would be complete by the time the disk had dispersed.

At a distance of 10^7 km the disk density will have fallen to 0.67 percent of its central value and we take this as the effective radius of the disk although, in principle, it stretches to infinity. The total mass of the disk, both gas and dust, beyond 4×10^6 km is over 8×10^{26} kg and if just 0.5 percent of that is deposited dust, assuming that not all is deposited, then the total mass of the carpet would be 4×10^{24} kg. Since the mass of the large Galilean satellites of Jupiter together have a mass about 4×10^{23} kg, there is an ample source of solid material to form the satellites.

Now we move on to the break-up of the dust carpet to give satellitesimals. The theory for the gravitational instability of a dust carpet was given by Goldreich and Ward in 1973.[4] A necessary condition for a satellitesimal to form at any location is that it must be able to withstand disruption due to planetary-produced tidal effects. The condition for this is (Steven writes)

$$\rho_B \geq \frac{3M_P}{2\pi R^3}$$

in which ρ_B is the density of the spherical blob, M_P the mass of the planet and R the distance to the centre of the planet. As the dust carpet is forming it is becoming denser and once the dust carpet reaches a thickness, h, such that its density reaches the critical level then gravitational instability will set in and the carpet will begin to break up. If the density per unit area of the dust component of

---

[4]Goldreich, P. and Ward, W.R. (1973) *Astrophysical Journal*, **183**, 1051–1061.

the disk is ρ_{ad}, then that thickness is given by (Steven writes)

$$h = \frac{\rho_{ad}}{\rho_B}.$$

According to Safronov,[5] the area of the disk in each of the condensations will be about $60h^2$ so the total volume of a condensation, which will form a satellitesimal, is about $60h^3$ with a mass, $m_B = 60h^3\rho_B$. For the parameters that gave the last figure I showed, the masses of satellitesimals as a function of distance from the planet are as shown here (Figure 8.5).

Jupiter has 63 satellites, a system dominated by the Galilean satellites, Io, Europa, Ganymede and Callisto, with masses $8.93 \times 10^{22}, 4.88 \times 10^{22}, 1.497 \times 10^{23}$ and 1.068×10^{23} kg, respectively. At a distance of 4.0×10^6 km, where settling times are within the expected disk lifetime, a satellitesimal mass is about 2×10^{22} kg, which is between about one-half to one-seventh the mass of the Gallilean

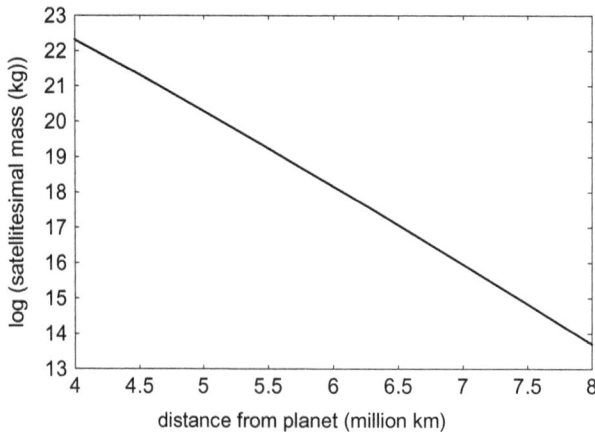

Figure 8.5   Satellitesimal masses at various distances from the planet

---

[5]Safronov, V.S. (1972) *Evolution of the Protoplanetary Cloud and Formation of the Earth and Planets* (Israel Program for Scientific Translation, Jerusalem).

satellites. The next most massive satellite of Jupiter is Amalthea, closer in than the Galileans and with a mass of 2×10^{18} kg within the range of masses of satellitesimals shown in the figure.

Now we come to the aggregation of satellitesimals to give satellites and it is here we shall see why the Safronov process can work for satellites but gives timescale problems for planets. The expression found by Safronov for the time to form a planet, adapted to satellite formation, is as I show here (Steven produces a pre-prepared written-out equation)

$$\tau_S = \frac{4r_L \rho_m P}{3\pi \rho_{ad}(1 + \beta)},$$

where r_L is the radius of the satellite, P is the period of a circular satellite orbit in the region of formation, β is a constant somewhere in the range 4–10, ρ_m is the density of the body being formed and ρ_{ad} is mean areal density of satellitesimals.

We found that for the model disk being considered here it was necessary to go out to a distance of at least 4.0×10^6 km to have a dust carpet formed in a sufficiently short time but the orbital radius of Callisto, the outermost Galilean satellite, is only 1.88×10^6 km. The answer to this problem is that protosatellite orbits are constantly decaying as the satellite grows and will still continue to decay for the duration of the planetary disk. This figure (Figure 8.6) shows the decay of a partially formed satellite, with constant mass 10^{22} km, moving in the disk that gave this previous figure (Figure 8.4) but with the disk density falling everywhere by a factor of e every million years. Starting with a satellite orbital radius of 7.5×10^6 km the orbit decays to the orbit of Callisto, the outermost Galilean satellite, in about 3×10^5 years and to the orbit of Io, the innermost one, in about 8×10^5 years. The orbit is close to circular for the whole of the decay period. Orbital decay is

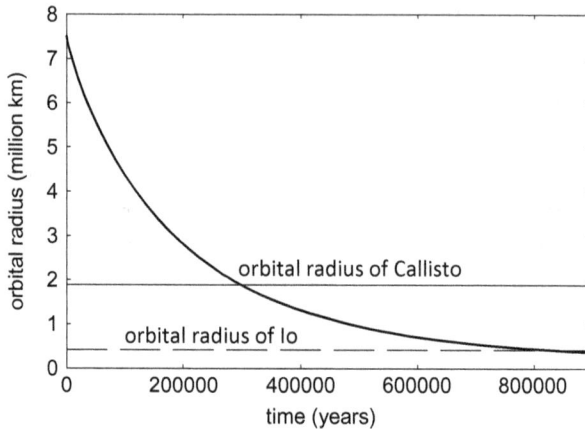

Figure 8.6   The decay of the orbit of a satellitesimal of mass 10^{22} kg

much faster than we previously found for planetary orbits because the disk density is much larger.

To see why formation times are so much shorter for satellites than for producing Jupiter according to the unmodified Safronov process, the main quantities we shall consider are the values of P and ρ_{ad}. We now do a very rough calculation. The period of Jupiter's orbit is approximately 12 years while that of Callisto, the outermost Galilean satellite, say formed at a distance of 7.5×10^6 km from Jupiter, is 130 days, a factor of about 34 less. The solar-nebula disk with a mass of, say, 50 Jupiter masses is spread out over an area of radius 35 au whereas the Jupiter disk with mass about one Jupiter mass is spread out over an area of radius, at most, 1 au, but probably much less. This gives another factor of 25 so that the estimated time for forming Jupiter due to these factors alone is of order 34×25 or 850 times as long as for forming Callisto. Allowing for the fact that $r_L \rho_m$ will be much smaller for a satellite than for a Jupiter core the time-difference factor will be at least 2,000. If it takes 10^8 years to produce Jupiter it would only take 50,000 years to produce Callisto. The calculation is very rough but makes clear the difference of timescales between producing planets and producing satellites by the basic Safronov process.

Last month I described how the commensurabilities were produced for the orbital periods the pairs of planets Jupiter-and-Saturn and Uranus-and-Neptune due to gravitational linkages between them while their orbits were decaying in a resisting medium. Commensurabilities also occur in the satellite systems of Jupiter and Saturn. The periods of the innermost Galilean satellites — Io, Europa and Ganymede — are in the ratios 1:2:4 and there are commensurabilities between various pairs of Saturn's satellites although, curiously, not necessarily neighbouring ones. The mechanisms for these commensurabilities will be exactly the same as for those of the planets.

That is about as much as I can say about satellite formation.

Solomon: That is fine Steve — I found your description quite convincing, especially as Simon was nodding his approval from time to time, although I think that he would contest your timescale for producing Jupiter.

Simon: I certainly would, although the conclusion that satellites form very quickly I do not dispute.

Solomon: Now let's move on. Recently I saw a science programme on BBC television that described a fairly complicated theory that gave Jupiter a major role in shaping the Solar System as it is today and also produced many other features of the system. It was all set against the background of the Nebula Theory and I must admit that I was a bit confused at times. Did you see the programme, Simon?

Simon: Yes, I did and they were outlining, for the benefit of non-scientific viewers, what is known as the Nice Model. Would you like me to go through this next week. It is not all that complicated and much easier to explain to a scientist than to television viewers.

Solomon: Next week is Easter weekend and I am off to stay with Alice's parents over the weekend. Her father is the local vicar — which sounds as though he might be dull but actually he is a very nice man with an interest in archaeology and I enjoy his company. So, if you agree, we meet again on 1st April, All Fool's Day; I hope that will not apply to you Simon.

Simon: April 1st is fine and I think that you will find that I have an interesting tale to tell and I will not at all look foolish.

Steven: So we meet again in a fortnight — we had better tell the landlord so he does not reserve our table next week.

They finish their drinks and as they leave Solomon informs the landlord that they will not be coming the following week but will come a fortnight hence.

Simon Describes the Nice Model and Steven Criticizes It

On 1ˢᵗ April 2016, Simon and Solomon are sitting in their reserved cubicle. Steven comes to the table, accompanied by the landlord carrying three glasses of Grant's Special.

Landlord: Welcome back gentlemen — I hope that you all had a good Easter break.

Solomon: Yes, thank you landlord, I had a fabulous time.

Steven: I just lazed around and caught up on my reading.

Simon: Lucky you; I took 3 days off and worked the rest of the time.

Landlord: I'll leave you now. I expect you want to get on with your discussions.

Solomon: Well Simon, you worked through some of the vacation and you're going to do some more work now. Are you ready to describe the Nice Model? If so then I hope that your explanation will be clearer than that I heard in the television programme.

Simon: Yes — we agreed that today I will describe the basics of the Nice Model. It shows how the major planets evolved to their present positions from a configuration where they were all much closer to the Sun than they are now. It also explains a number of other features of the Solar System and events that occurred during its evolution.

Solomon: Sorry to interrupt, Simon, does the name 'Nice model' indicate an origin in the south of France?

    Simon: Yes it does, the model was produced by four astronomers — Rodney Gomes, Hal Levison, Alessandro Morbidelli and Kleomenis Tsiganis — based at the Observatoire de la Côte d'Azur in Nice.[1]

    Their model is based on simulations of the evolution of the orbits of the four major planets, starting from when the gas disk had disappeared but in the presence of a dense planetesimal disk, of total mass between 30 and 50 Earth masses, that has an inner boundary just outside the initial region of planetary occupation and stretches out to between 30 and 35 au. The areal density of planetesimals was taken to fall inversely with heliocentric distance and was sharply terminated at the outer limit.

    It is assumed that any planetesimals within the region of the giant planets would either have been absorbed by the planets or have been gravitationally ejected from the region. In all the simulations, the major planets were initially in close-to-circular orbits with Jupiter's orbit of radius 5.54 au and Saturn's of radius 8.65 au. This puts Saturn just inside the distance at which its orbital period would be twice that of Jupiter. The outer ice-giant planets were placed at different distances in different simulations with the radius of the Neptune orbit between 11 and 13 au and that of Uranus between 13.5 and 17 au but with the proviso that the difference of their orbital radii should be more than 2 au. You will notice that, initially, Neptune is closer to the Sun than Uranus is, unlike the present position. All the orbits were very close to coplanar. The chosen total mass of the planetesimals was represented by anything between 1,000 and 5,000 bodies of equal mass. In a real situation, the actual

---

[1]Tsiganis, K., Gomes, R., Morbidelli, A.F. and Levison, H. (2005) *Nature*, **435**, 459–461.

number of planetesimals would be much larger. Assuming that an average planetesimal had a mass of 10^{19} kg, corresponding to a radius of about 100 km, then there would be something of order 20 million, or more of them, but the statistical behaviour of the system should have been adequately reproduced by the number in the simulation.

A few weeks ago, I showed you this figure (Figure 4.1) illustrating the interaction of a planet on a circular orbit with both inner and outer planetesimals on originally-circular orbits. It showed that the interaction with an inner planetesimal moved the planet outwards because it increased its tangential component of velocity and hence increased its angular momentum. Conversely, interaction with an outer planetesimal reduced the angular momentum of a planet and moved it inwards. Since the areal density of planetesimals in the simulation decreases outwards the net result of planetesimal interactions is to move the planets outwards, an analytical prediction that was confirmed by the computer simulations. In the simulations, the planets interacted gravitationally with all other bodies but the planetesimals did not interact with each other. Forty-three different scenarios were run, with a wide variety of outcomes, but some general patterns of behaviour became evident.

A typical output is shown in this figure (Figure 9.1). Three lines are shown for each planet. The top one gives the aphelion, Q, i.e. the furthest distance of the planet from the Sun, in au. The bottom line shows the perihelion, q, the closest distance and the middle line is the average of the two, a, known as the semi-major axis. The difference between Q and q is indicative of the eccentricity, e, of the orbit, i.e. $e = (Q-q)/(Q+q)$ — shown on the right-hand side adjacent to the double-arrowed line.

It will be seen that, as previously indicated, the starting point gives Neptune closer than Uranus to the Sun. For about 6 million years, there is a very slow, almost

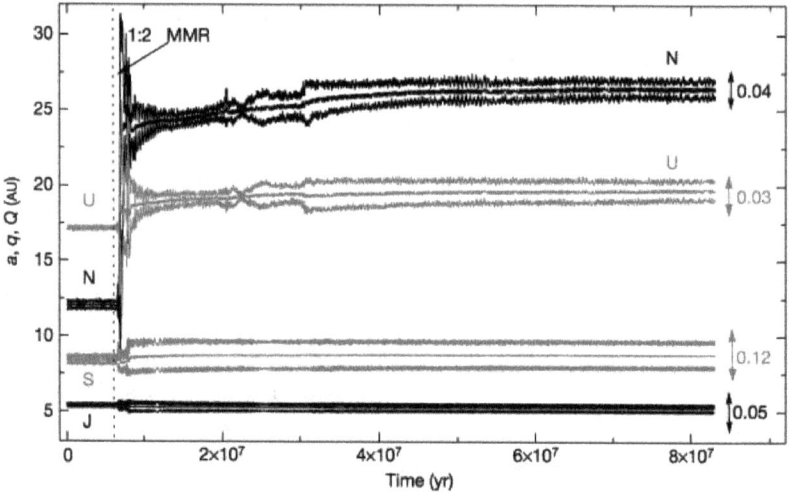

Figure 9.1   The evolution of the orbits of the giant planets in one run of the Nice model

imperceptible, evolution of the orbits; the planetesimals are slightly affecting the ice-giant planets but the
main variations are due to the gravitational interactions
between the planets themselves. Then the Jupiter and
Saturn orbits reach a 1:2 ratio in their orbital periods,
shown as MMR (mean motion resonance) in the figure.
When this happens Jupiter and Saturn repeatedly come
closest together at the same points in their orbits so their
mutual perturbation is enhanced and the orbit of Saturn,
the less massive planet, is most greatly affected. Its orbital
eccentricity is increased bringing it into close proximity to
Neptune and, in an interaction between them, Neptune is
propelled outwards into a highly eccentric orbit beyond
the orbit of Uranus. Uranus is also perturbed into a
more eccentric orbit and now both Uranus and Neptune
have penetrated into the planetesimal region. Due to
the interaction of the planets with the planetesimals,
through a process known as dynamical friction, which
is similar in its action to that of a gaseous resisting

medium as previously described by Steve, in time the outer-planet orbits become more circular. When the orbits of the ice giants have rounded-off within the planetesimal region they are further influenced by interactions with the planetesimals, the net result of which is to give a gradual drift outwards.

Many planetesimals are ejected inwards and their entry into the Jupiter–Saturn region gives a sudden increase in the rate of orbital evolution of those planets. After 80 million years, it can be seen that the system has settled down. The eccentricities are of the same order as those of the present giant planets, with the exception of Saturn, the indicated eccentricity being 0.12 whereas the present eccentricity is 0.056.

Each run of the Nice Model gave a different outcome but patterns emerged giving particular types of outcome depending on whether or not certain events took place. There were two types of run of the model that could be regarded as 'unsuccessful' in the sense that the outcomes were very different from the arrangement of giant planets that now exists. The first of these, in which the initial difference of the semi-major axes of Saturn and Neptune was more than 5 au, gave no interactions between major planets and just slow orbital evolution occurring over an extended period. Uranus remained the outer planet with its final orbital radius less than 16 au. These runs corresponded to the most extended starting configurations. The other unsuccessful runs were those in which the initial difference in the semi-major axes of Saturn and Neptune was less than 3 au, in which Neptune was scattered inwards by an interaction with Saturn and subsequently interacted with Jupiter and was ejected from the Solar System.

There were 29 'successful' runs, giving outcomes that could be related to the present arrangement of giant planets, which the Nice workers divided into two classes.

Class A consisted of runs, in which only the ice giants interacted and class B in which Saturn interacted with one or both of the ice giants. Both classes gave final outcomes having planetary orbits in the presently-observed order of distances from the Sun and with small eccentricities and inclinations. On the whole, the class B outcomes could much more closely be identified with the characteristics of the present Solar System.

This figure (Figure 9.2) summarizes the outcomes for the successful runs. I'll now take you through an explanation of what it shows. For each of the classes,

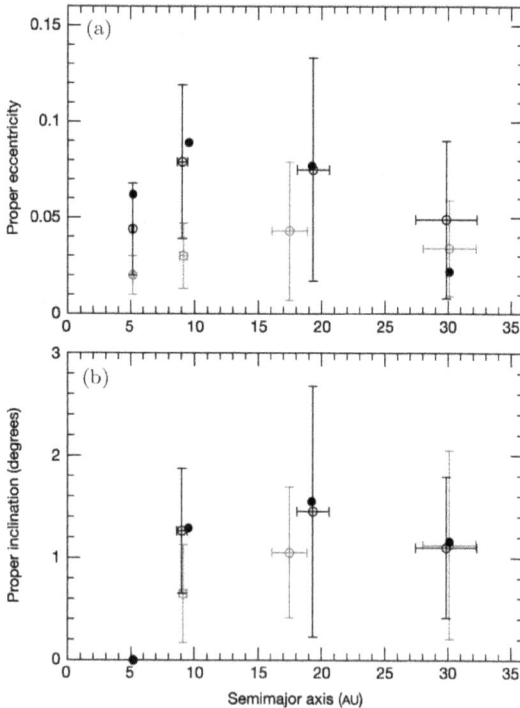

Figure 9.2   The final outcomes for average semi-major axes, eccentricities and inclinations, and their standard deviations, of class A (grey open circles) and class B (black open circles) simulations. The true values of semi-major axis, eccentricity and inclination are shown by the solid black circles (Tsiganis *et al.*, 2005)

the averages of the eccentricities, orbital inclinations and semi-major axes were found together with the standard deviation of the values for each quantity. The class A results are shown as the open grey circles and the class B as the open black circles. The true values are indicated by the solid black circles. The inclinations are measured relative to that of Jupiter — hence Jupiter's inclination is taken as zero. Normally, solar-system planet inclinations are given relative to the ecliptic, the plane of the Earth's orbit, in terms of which the inclination of Jupiter is just over 1°.

As will be seen, the results for class A (15 runs) and class B (14 runs) are both quite good with class B giving closer agreement, especially for the semi-major axis of Uranus. The vertical and horizontal bars give the standard deviation for each quantity for each class and it will be seen that most averages are within one standard deviation of the true value.

One possible problem with the model was investigated — the survivability of satellites of the ice giants due to the various interactions that take place. Eight of the simulations gave approaches of Saturn close enough to an ice giant to suggest that the satellite systems of the planets should be disturbed and possibly disrupted. These runs were repeated with the satellites included around the ice giants, with both being given the present satellites of Uranus. In one half of the encounters all the regular satellites — those orbiting closest to the planet in circular orbits in the planet's equatorial plane — survived but outer irregular satellites, identified as captured objects, were lost. This means that the irregular outer satellites would have to have been captured after the close interactions.

The Nice Model seems to explain well, in a general sense, the present configuration of the major planets. Of course, no individual simulation gives very close agreement with the present configuration but with such a complex system it is impossible to analyse every possible initial configuration. This model has a very high level of acceptance by planetary scientists.

I think that I would now like to have a short break to quench my thirst. Drink up you two and I'll get some refills.

Solomon: This seems a good time to take a break. I have been quite impressed by the Nice Model and I can see why it has had such wide acceptance. It seems to explain the arrangement of major planets rather well.

The three chat while they finish their drinks. The Simon goes to the bar and returns will the recharged glasses.

Solomon: Are you ready to carry on, Simon? I gather that there is more to come from the Nice model.

Simon: Yes, indeed, there are a number of other solar-system features that can be explained, all concerned with the final destinations of planetesimals.

The study of craters, primarily on the Moon, indicates that about 500 million years after its formation, bodies in the inner Solar System were subjected to a period of heavy bombardment. Once Uranus and Neptune penetrated the planetesimal region then many of them were perturbed to move inwards, which is, after all, the mechanism previously described for motions of the outer planetesimals when they interacted with the icy planets. This rain of planetesimals, penetrating into the inner Solar System, known as the Late Heavy Bombardment, usually referred to as LHB, led to the cratering not only of the Moon but also of all other bodies in the terrestrial region. While this bombardment also happened to the Earth, because of plate tectonics and constant renewal of the Earth's surface, all evidence of it has been obliterated.

The inward rain of planetesimals can explain many other solar-system characteristics. These planetesimals will not only collide with planets but also interact with them gravitationally. The combination of this and collisions of pairs of planetesimals would result in them ending up in a large variety of orbits. Some finished in the Asteroid Belt between Mars and Jupiter, keeping well away from those planets, and are still there today. Although there are many bodies in the Asteroid Belt their total mass is just 4 percent of that of the Moon so they are just that tiny fraction of the total mass of planetesimals that happened to end up in 'safe' orbits.

Another feature of the simulations was that it was found that some of the planetesimals ended up as so-called Trojan asteroids that have been observed to accompany Neptune and Jupiter. These are bodies that have virtually the same orbits as the planets but are either 60° ahead or 60° behind the planet in its orbit. These are regions where a body is in a stable configuration under the combined gravitational influence of the Sun and the planet. They are stable in the sense that if for any reason they depart from the point of stability the combined forces of the Sun and planet push them back to where they came from. In practice, there can be a number of Trojan bodies for each planet and they will oscillate to-and-fro around the point of stability. Their numbers are few and again they can be thought of as a tiny fraction of the planetesimal population that became trapped in these positions.

The original spread of planetesimals extended beyond the present orbit of Neptune, When Neptune migrated out to its present orbit at a distance of 30 au from the Sun it influenced the planetesimals in its vicinity outside its orbit. Some were scattered inwards, those passing close to the Sun appearing as comets, which make themselves visible by the vaporization of their volatile content due to heating by the Sun. This still happens, although at a lower rate — about once every hundred years — to produce what are known as *short-period comets*, which spend a large part of their existence within the planetary region. After about 1,000 excursions close to the Sun their volatile content is exhausted and then they become dark inert objects, something like low-density asteroids. A well-known short-period comet is Halley's Comet, with a period of about 75 years.

Other planetesimals were scattered outwards by Neptune to form the Kuiper Belt, named after the Dutch, later American, astronomer Gerard Kuiper. He wrote an article in 1951 speculating on the existence of a belt of asteroids beyond Neptune, although others had predicted its existence much earlier. This belt, as so far discovered, seems to have about 20 times the width and 200 times the mass of the Asteroid Belt.

There is a class of objects known as dwarf planets that are in heliocentric orbits, have insufficient mass to clear all objects close to their orbits but are massive enough to have formed themselves into spherical, or near-spherical, bodies under the influence of their self-gravitational forces. Five of them are in the Kuiper Belt — Pluto, Eris, Haumea, Makemake and V774104, the last having been discovered at the end of 2015, and a sixth is Ceres, a body of diameter roughly 1,000 km that resides in the Asteroid Belt. Dwarf planets represent collections of planetesimals that never progressed to a size that would qualify them to be fully-fledged planets. If their original orbits were outside that of Neptune then, by interactions with that planet, they would have been pushed further outwards.

An interesting feature of orbital evolution is that some Kuiper-belt objects are in orbital resonance with Neptune, meaning that the orbital periods of the two bodies are related as the ratio of two small integers. The outstanding example is Pluto, the average orbital period of which is related to that of Neptune by the ratio 3:2. The actual present ratio is 1.504 but Pluto's orbit slightly changes with time and the time-average of the ratio is 1.5.

Although Pluto's orbit at perihelion is just inside the orbit of Neptune the bodies never approach closely, partly because of the inclination of Pluto's orbit, 17°, but mainly because there are dynamic processes that keep them apart as well as maintaining the 3:2 orbital resonance. Actually Pluto approaches Uranus more closely than it approaches Neptune.

There is another source of comets, the Oort Cloud situated at tens of thousands of au from the Sun and stretching out a long way towards the nearest stars. The bodies in the Oort Cloud were produced early in the formation of the Solar System by interactions of planetesimals with Jupiter. The occasional passage of stars close to, or through, the Oort Cloud perturbs cometary bodies to move inwards and if they move close enough to the Sun they become visible comets. Because they come from so far they have eccentricities close to unity — all elliptical orbits must have an eccentricity less than unity — and large semi-major axes so they are very weakly bound to the Solar System. The energy of an orbiting body consists of two parts, kinetic energy due to its motion, which is positive, and potential energy due to its distance from the Sun, which is negative. If the total energy — the sum of the two — is negative then the orbiting body is bound to the Solar System and will be in an elliptical orbit. However, if the total energy is positive then the body will escape from the Solar System. For comets coming from the Oort Cloud the two elements of the total energy are almost in balance but slightly negative. In passing through the region of the planets the comets gain or lose an amount of energy by planetary perturbations with the expected magnitude of the gain or loss greater than that of the initial total negative energy of their orbits. If they gain energy then their total energy will become positive and they will eventually escape from the Solar System. If they lose energy then their new negative energy will be proportionately much greater in magnitude than previously and they will go into orbits with much smaller semi-major axes — the semi-major axis is proportional to the inverse of the total energy. This means that either they will never appear again or, if they do so, they will appear in a very different orbit. A consequence of this scenario is that it is almost certain that they have never been seen before as objects coming from the Oort Cloud and for this reason they are called *new comets*.

That completes my summary of the Nice model. As I said previously, it explains a great deal of what we know about the Solar System and for that reason it has very wide acceptance.

Solomon: I find it pretty convincing myself — what do you think Steve?

Steven: The simulations are pretty good and, as Simon has indicated, it would be too much to expect that any one of them would give anything very close to what we have today. There are too many parameters to vary to explore all possibilities. If the Nice model is to be considered as part of the Nebula Theory then my main criticism concerns the starting point for the simulation, with the four major planets strung out between about 5 and 17 au. On 6th February, we had a meeting in which the formation times for terrestrial planets and the cores of major planets were discussed. You will recall that, according to the basic Safronov theory of planetesimal accumulation, forming a Jupiter core in the present vicinity of Jupiter would take of order 10^8 years and much longer further out. A sticking-plaster solution for this — the runaway-growth model — was suggested by Wetherill and Stewart but even this would not allow a major planet to form at 17 au from the Sun. Again, Wetherill himself pointed out a problem with the runaway model that, because of planetesimal scattering by the growing cores, the basic assumption of the runaway model, that the relative speeds of the cores and local planetsimals should be small, will not be satisfied. More recent work in 2000 by two Japanese astronomers, Eiichiro Kokubo and Shigeru Ida,[2] simulated planet formation from planetesimals, incorporating the concept of runaway growth and including the effect of gas drag and came up with formation times for the Earth, Jupiter, Saturn, Uranus and Neptune in their present locations of 0.7, 40, 300, 2,000 and 7,000 million years, respectively. They put every factor that could operate in producing the giant planets and came up with those figures for producing them in their present locations. From their results it is clear that to produce major planets within the probable maximum lifetime of the gaseous circumstellar disk — 10 million years — you need to produce them all well within the present orbit of Jupiter. How then do you get the starting point for the Nice model? You cannot just brush this problem under the carpet and go on to further theoretical developments like the Nice model on the basis that the starting point is not a problem. If you can get to the starting point of the Nice model then it works but there has been no convincing demonstration that the starting point is realistic.

There are other problems but of a lesser nature. For example, Uranus and Neptune enter the planetesimal region after a few million years and begin to perturb planetesimals inward from that time. We are also told that after 80 million years the system had settled down. There is then the question of why it was that the LHB occurred 500 million years after the formation of the Solar System. The gas component of the nebula would have disappeared after 10 million years at most so what is happening in the almost 500 million years to the beginning of the LHB? Is this time taken up with creating the starting point for

---

[2]Kokubo, E. and Ida, S. (2000) *A New Era in Bioastronomy*, ASP Conference Series, Vol. 213, Eds. G. Lamarchand and K. Meech.

the Nice Model originating with giant planets much further in? If so then how do you progress to the starting point both for the positions of giant planets and the distribution of planetesimals assumed by the Nice Model?

You will have noticed the importance of the 2:1 Saturn–Jupiter resonance for the Nice Model. However, the actual present resonance is 5:2, so how does it get there? I described it as due to Saturn–Jupiter interactions in a gaseous medium. Can the same happen in a medium consisting of planetesimals and, if so, how? Do any of the simulations give a 5:2 resonance or anything close to it?

A final minor point concerns the figure Simon showed us that marked the true eccentricities of the major planets as solid black circles. That figure comes from a paper written by the Nice astronomers but the eccentricities they indicate are completely wrong. The true values from Jupiter outwards are 0.0484, 0.0542, 0.0472 and 0.0086, nothing like what is shown in the figure. Am I missing something here?

Solomon: Any answer to this Simon?

Simon: I have a feeling that early events in nebula development are extremely complex and that it is difficult to take all factors into account, even if we knew what all the factors were. The mass, size and distribution of the original nebula, plus the nature and composition of the dust within it, are all unknowns and will critically affect the behaviour of the system. We make simple models of complex systems but we have to accept their limitations.

Steve has described the Capture Theory up to the level of producing planets, both terrestrial and major, together with satellites, but can he offer any explanations for the various features of the Solar System that the Nice model explains?

Steven: Yes I can, and I suggest that is what we should concentrate on next week.

Solomon: Despite what you say Steve I think that the Nice Model has considerable merit and I would be interested to hear what you have to say and whether it could be equally or more convincing.

They finish their drinks while tell each other of their activities over Easter. Solomon announces that he and Alice are to be married in October with Alice's father officiating.

The Tenth Meeting

Steven Describes the Formation of Asteroids and Comets and the Origin of Many Other Features of the Solar System

On 8th April 2016, the friends are sitting in their cubical and the landlord approaches with three brimming glasses of Grant's Special that Solomon had previously ordered. After some preliminary chat they begin their formal proceedings.

Solomon: Steve, I believe you are going to tell us how the formation of asteroids, comets and various other small bodies are explained in the context of the Capture Theory. Will what you tell us be as convincing as what we heard last week, I wonder?

Steven: I hope so, but that will be for you to judge after I have presented the ideas about their formation. What I shall be describing all follows from the planetary collision I described three weeks ago, the simulation of which is shown in this figure (Figure 7.1). The total mass of iron-plus-silicate in Bellona and Enyo was 27.5 Earth-masses but the residual cores of the colliding planets that formed the Earth and Venus contained only a small part of that inventory of iron and silicate, the remainder being in the form of debris. In addition, there is a great deal of light-atom molecular material that condensed into ices when it cooled and formed either totally icy bodies or became

159

part of other bodies that formed. Some of it would have become incorporated as a tiny component of the surviving cores and have given the water content of the Earth and possibly some of the methane that is now being extracted by fracking all over the world or is trapped in the frozen soil of tundra regions — although most methane is taken as being of biological origin.

The figure shows that debris is thrown out in all directions; the final destination of the debris depends on its intrinsic energy — the energy per unit mass, part kinetic and part potential as Simon explained last week. If this is negative then the debris will be retained but if positive it will be expelled from the Solar System. The intrinsic energy was found separately for core, mantle and ice material for those SPH particles outside the spheres of influence of the two cores — the regions close to the cores where their gravitational influence is dominant. This figure (Figure 10.1) shows the mass in Earth-units per unit intrinsic energy, expressed in $GJ\,kg^{-1}$ (see footnote 1) for each of the components. More than one-half of the core material is retained and a somewhat smaller fraction of mantle material. However, the majority of the ice is expelled from the system. This pattern of material loss is expected from the proposed structures of the colliding planets. Material further from the centre of the planets was less constrained in its motion when the nuclear explosion occurred, so that it will have moved further from the collision and a greater proportion would have been lost. Although the planets were modelled with sharp divisions between the different layers, in practice there will be gradual changes in composition. Thus, the further out the mantle silicate was, the higher would have been its volatile content. From deduced compositions of asteroids,

---

[1]$GJ\,kg^{-1}$ indicates 10^9 joules per kilogram where the joule is the basic unit of energy in Standard International (SI) units.

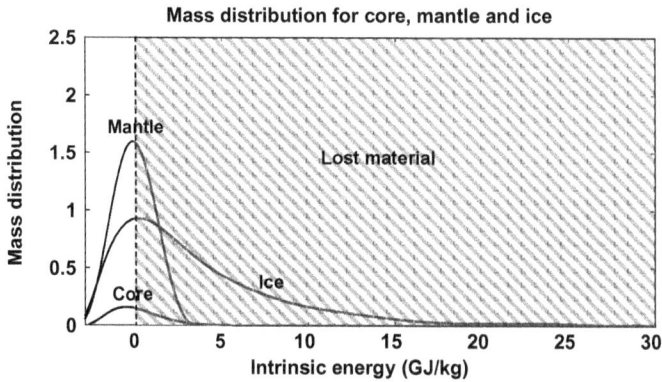

Figure 10.1   The distribution of dispersed material from the collision. The graphs show the distribution of mass in Earth-units per unit intrinsic energy (GJ kg^{-1}). The shaded region corresponds to material escaping from the Solar System

as judged by the spectral features of reflected sunlight, C-type asteroids, containing the most volatile material, tend to be further from the Sun than S and M-type asteroids, associated with silicate and metal compositions — which is consistent with the model of debris distribution suggested here. From the areas under the curves in the figure it is clear that by far the most abundant retained material is silicate, followed by ice and then iron. We identify this debris as the source material of asteroids and comets, where comets are defined as bodies containing a considerable proportion of volatile material.

Meteorites are fragments from colliding asteroids that fall to Earth. Since stony meteorites, those mostly consisting of silicates, are similar to Earth rocks in composition and, once weathered, similar in appearance they are more difficult to identify than iron meteorites, dense lumps of iron plus some nickel. However, sometimes meteorites are seen to fall to Earth, events known as *falls*, and, because they have not been exposed to weathering, they are readily recognized as meteorites when found, regardless of their type. Other meteorites are found in Antarctica on or near the surface of ice sheets several kilometres thick so it is clear that they came from above and not from below. For falls and Antarctic meteorites, which are regarded as an unbiased sample of asteroid material, the ratio of stones to irons is about 29:1. A large ratio of silicate to iron is consistent with what we see in the figure.

The feature of meteorites that ties up with a planetary origin is their compositions. The great majority of meteorites contain both iron and silicate but usually one or other component dominates so we can refer to them as either stones or irons. These can be identified as coming from different regions of a planet in which differentiation by density was well advanced but not quite complete. There are a few meteorites, about 1 percent of the total, that contain substantial amounts of both components that are called stony-irons. One type of stony iron, *pallasites*, shows globules of silicates, mainly the mineral olivine, within an iron matrix, as seen here (Figure 10.2). This can be interpreted as silicate trapped in an iron core with globules rising through the metal ever slower as the iron cooled until eventually they were trapped in solid metal.

Figure 10.2   A Pallasite with olivine crystals (dark) within a metal matrix (Oliver Schwartzbach)

The other kind of stony-iron, *mesosiderites*, is quite different. The rock is partially fragmented and the mineral components are only stable at low pressures, so they could not have originated deep in a planet. The metal is partly in the form of globules and partly as veins occupying space between minerals. They are quite rare and it is thought that they might have their origin in asteroid collisions.

The idea that asteroids came from a shattered planet goes back some time. It was early realized that they were of irregular shape; as they tumble in space so the amount of light they reflect towards Earth varies and they fluctuate in brightness. The most likely source seemed to be the debris of a broken planet and in 1973 Napier and Dodds expressed this idea in a scientific paper entitled 'The missing planet'.[2] However, nobody could envisage any source of energy that could break up a planet and widely disperse the debris so the idea fell out of favour The capture-theory model, and the subsequent planetary collision, with D–D reactions providing a copious source of energy, provides the mechanism by which planets *can* break up and provide fragments.

Later, when I have finished our talk it might be interesting to be told how meteorite, and hence asteroid, compositions are explained by the Nebula Theory. There seems to be no reason why the planetesimals formed by the Goldreich and Ward mechanism should not be just intimate mixtures of iron and silicate grains. I know what the nebula-theory explanation is but you, Sol, will be able to compare the two different explanations from the two theories.

The eccentricities and inclinations of the orbits of the retained material show some interesting trends, as seen in this figure (Figure 10.3). Observations show that nearly

---

[2]Napier, W.M.D. and Dodd, R.J. (1973) *Nature*, **224**, 250–251.

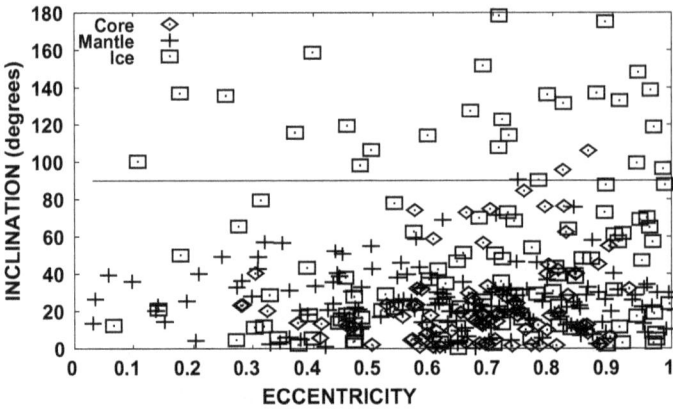

Figure 10.3   The eccentricities and inclinations of the retained material (Woolfson, 2013)

all the iron and silicate orbits are prograde, i.e. have inclinations less than 90°, and hence are orbiting in the same sense as the Sun's spin, and the majority of these have inclinations less than 45°, which is what the figure shows. There are a few asteroids and rather more comets that have been observed in retrograde orbits, for example Halley's Comet, but a considerable proportion of the ice from the model is in retrograde orbits. However, bodies in retrograde orbits in the presence of much larger numbers of bodies in direct orbits are likely to collide and be fragmented and so be less likely to be observed.

Asteroids, formed from debris, would have interacted with each other and with planets, the orbits of which were still evolving, so that interactions could have occurred at large distances from the Sun. Some would have attained safe orbits, such as those in the Asteroid Belt, mainly between Mars and Jupiter, or in the region between Saturn and Uranus, now occupied by the large asteroid Chiron — which is said to have comet-like characteristics — and other bodies. The total mass of the surviving asteroids in the Asteroid Belt, about 4 percent of that of the Moon, is small compared with the amount of material originally present since most of it would have been swept up by the major planets once they settled into their final orbits.

The source of observed comets is associated with two regions of the Solar System — the Kuiper Belt, just beyond the orbit of Neptune, and the Oort Cloud, tens of thousands of au from the Sun and stretching out almost half way to the nearest star. It has been suggested that Oort-cloud comets have an origin from outside the Solar System because they have D/H ratios 20 times the cosmic value, but the high D/H value that we have taken as occurring in the ice of the colliding planets weakens that argument. In 1983, Mark Bailey proposed that there is an inner reservoir of comets, between the Oort Cloud and the Kuiper Belt, which is drawn outwards to replenish the Oort Cloud when it is depleted by a very severe disturbance, such as when a star passes near or through it or when the Solar System passes through a Giant Molecular Cloud, a huge cloud of molecular material with a mass typically 10^5 times that of the Sun.[3] The results in this figure (Figure 10.1) are consistent with the Bailey model since the intrinsic energies of both the mantle and ice debris give a continuous distribution of debris stretching from the inner Solar System to the region of the Oort Cloud. If the continuity of the distribution has persisted throughout the lifetime of the Solar System then the observed parts of the Kuiper Belt form the inner region of this distribution. Short-period comets, ones that spend most of their lives within the planetary region, are those from the Kuiper Belt that are perturbed by Neptune into orbits taking them close to the Sun so that they become visible. The so-called *new comets* that Simon described last week come from the Oort Cloud due to perturbation by some body external to the Solar System. Since there

---

[3]Bailey, M.E. (1983) *Monthly Notices of the Royal Astronomical Society*, **204**, 603–613.

are no major sources of perturbation for the comets at the centre of the distribution, which may be the majority of all comets, there is no way in which their presence has, so far, been detected.

A characteristic of orbits is that if the only forces at play are the gravitational forces of the two principal bodies, for example, the Sun and an asteroid, then they repeatedly go through the same regions of space. Hence, once the planetary orbits had stabilized, a debris fragment beginning in the terrestrial region will repeatedly pass through the region occupied by the planets and inevitably it will eventually collide with one of them or be thrown out of the Solar System. For asteroids not in safe zones between pairs of planets, only if its perihelion keeps it beyond Neptune will it have a permanent or at least long-term existence.

Some cometary material that went further out, could have interacted with major planets in evolving orbits at hundreds of au from the Sun and be swung into orbits well outside the present planetary region. With the notation aphelion $= Q$ and perihelion $= q$, this figure (Figure 10.4) shows the interaction of a comet with $(Q, q) = (110\,\text{au}, 0.5\,\text{au})$ with a Jupiter-mass planet with $(Q, q) = (100\,\text{au}, 10\,\text{au})$ with closest approach of the bodies 1.84×10^6 km. After the interaction, the comet orbit is given by $(Q, q) = (109.9\,\text{au}, 42.1\,\text{au})$, which would place the body well within the Kuiper Belt. This process is most efficient if the interacting bodies have similar aphelia so it is unlikely that many cometary fragments will be affected in this way.

Some weeks ago, I described the various ways that planetary orbits evolved within a resisting medium. The main mechanism is due to the planetary mass affecting the medium, with the resistance coming from the reaction on the planet. However, bodies as small as comets do not have sufficient mass greatly to influence the distribution of the medium and the resistance is primarily due to the

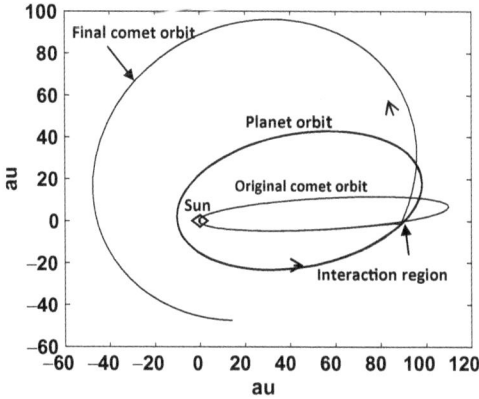

Figure 10.4   Debris interacting with a planet in an evolving orbit is perturbed into an orbit with a small perihelion to a much larger one (Woolfson, 2013)

ram pressure they experience because of the impact of the medium on the comet. This is similar to air resistance experienced by a body, such as a rider on a bicycle or a motor vehicle. For a spherical comet, the force experienced will be (Steven writes)

$$F = \pi \rho a^2 V^2,$$

where ρ is the local density of the medium, a the radius of the comet and V the velocity of the medium relative to the comet. The force is in the direction of the motion of the medium relative to the comet. The effect of such a force has been found for a comet of mass 7×10^{12} kg, of density 500 kg m^{-3} (published estimates are between 100 and 1,000 kg m^{-3}), giving a radius of 1.5 km, with original perihelion 0.5 au and various original semi-major axes. The medium had 40 times Jupiter mass with an annular distribution of density, like this (Figure 4.7) with a flared structure perpendicular to the mean plane. Mathematically, the density of the medium was given by (Steven produces a pre-prepared equation)

$$\rho(r, z) = C \exp \left(-\frac{(r - d)^2}{2\sigma_r^2} \right) \exp \left(-\frac{h r^2}{s^2} \frac{z^2}{\sigma_z^2} \right),$$

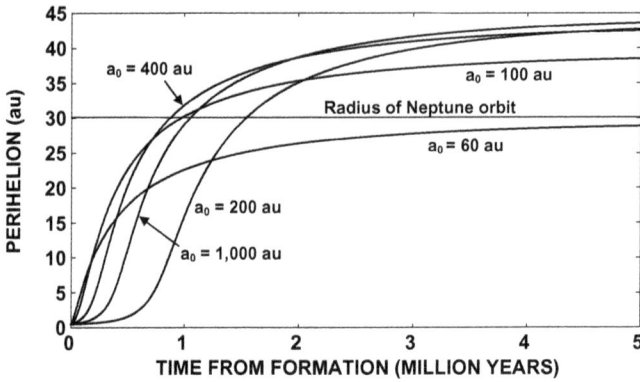

Figure 10.5   The changes of the perihelion with time for comets with original perihelia 0.5 au and various initial semi-major axes, a (Woolfson, 2013)

where $d = 100$ au, $\sigma_r = 30$ au, $h = 10$, $s = 20$ au, $\sigma_z = 30$ au and r and z (distance from the mean plane) are both expressed in au. The first exponential term describes the distribution in projection and the second gives the flared structure perpendicular to the mean plane. The constant C is determined from the total mass of the disk. The results are shown in this figure (Figure 10.5).

It will be seen from the figure that all orbits with original semi-major axes greater than about 70 au will end up with perihelia beyond the orbit of Neptune and within the Kuiper-belt region. This means that once the planets had settled down into their final orbits the comets were safe except for possible perturbation by Neptune. Their aphelia stretch out to several thousand au and represent the complete range of cometary occupation from the Kuiper Belt to the Oort Cloud, including the inner cloud of comets postulated by Bailey as a reservoir for the replenishment of the Oort Cloud.

I'll now pass on to something else but before I do so I think I would like a break. Drink up and I'll get refills.

Solomon: This seems a good time to take a break especially as I would like to look through my notes. Steve has given me plenty to think about.

Simon and Steven drink and chat for a few minutes while Solomon studies his notes. Then Steven went to the bar and returned with glasses replenished.

Solomon: Are you ready to continue, Steve?

Steven: Yes, I am. The topic I am now going to address concerns three bodies that are clearly related in some way and I am going to explain how that relationship could have come about.

Pluto, a dwarf planet, has an orbit of semi-major axis 39.54 au, eccentricity 0.249 and inclination 17° that passes just within the orbit of Neptune, although the bodies never approach closely. The periods of the orbits of Pluto and Neptune have a 3:2 commensurability and the times they are closest do not correspond to when Pluto is at perihelion. As Simon told us last week, rather curiously Pluto approaches Uranus more closely than it does Neptune. The third body is Triton, the seventh largest satellite in the Solar System that is in a retrograde orbit around Neptune, which rules it out as a regular satellite. Another peculiar satellite of Neptune is Nereid, in an extended direct orbit of eccentricity 0.75. In 1999, Woolfson explained the relationship between these bodies as just another outcome of the planetary collision.[4]

The scenario for this explanation is that Triton was a satellite of one of the colliding planets, released into an extended heliocentric orbit taking it well beyond the present orbit of Neptune. Neptune had a family of regular satellites, the largest of which was Pluto with mass about 60 percent that of Triton. A computer simulation was made of a collision involving Triton and Pluto. For the starting point shown here (Figure 10.6), Triton was in a direct heliocentric orbit with perihelion 2.6 au and aphelion 55.6 au. Pluto was in a direct circular orbit, of

---

[4]Woolfson, M.M. (1999) *Monthly Notices of the Royal Astronomical Society*, **304**, 195–198.

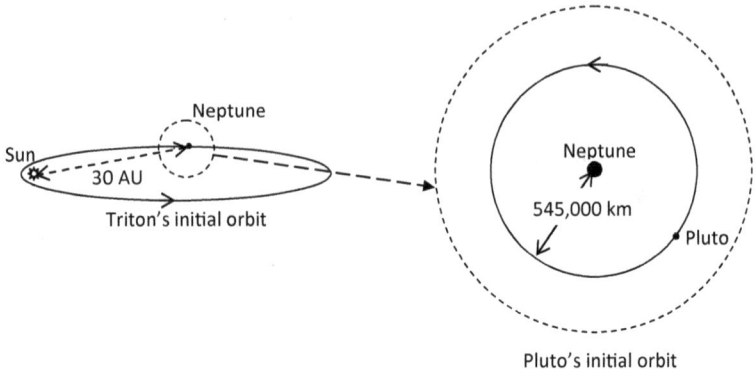

Figure 10.6   The initial orbits of Triton and Pluto before the collision

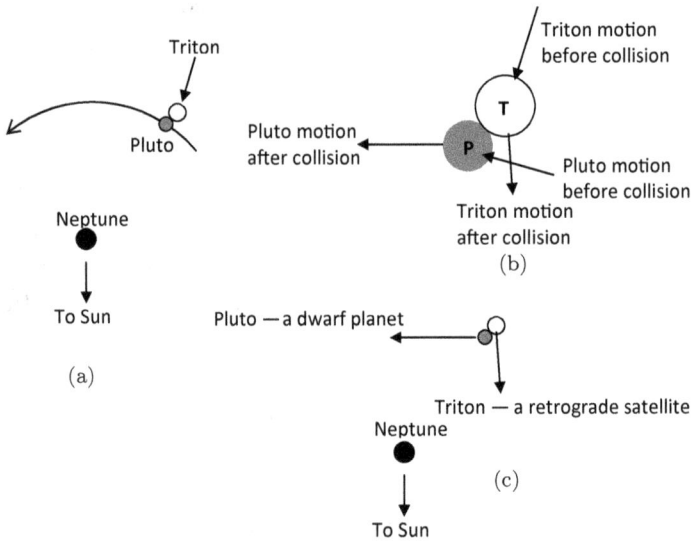

Figure 10.7   The Triton–Pluto collision. (a) Triton, travelling towards the Sun strikes Pluto; (b) motions before and after the collision; and (c) the final outcome

radius 545,000 km around Neptune. The before-and-after collision situations are shown in this figure (Figure 10.7). Triton, moving towards the Sun, struck Pluto a glancing blow that ejected it from its orbit around Neptune into a heliocentric orbit with $(a, e) = (39.5\,\mathrm{au}, 0.253)$, very similar to its present orbit. Because of the way it was struck by Triton, Pluto was set into retrograde spin and a large part of it was sheared off to form its comparatively large satellite Charon in a retrograde orbit. Fragments from the collision gave Pluto's four other much smaller satellites. Notice the similarity of the way that Charon's relationship to Pluto is established and that of the Moon to the Earth by the collision model, described by Simon.

Triton's speed relative to Neptune was reduced by the collision and it was captured by Neptune into a *retrograde* orbit with $(a, e) = (436,000 \text{ km}, 0.88)$. It was shown by McCord that retrograde satellite orbits quickly round-off and decay,[5] eventually giving Triton's present orbit with $(a, e) = (355,000 \text{ km}, 0.000)$.

The extreme nature of Nereid's orbit may be due to disturbance of a less eccentric orbit by Triton's incursion into the vicinity of Neptune. Alternatively, and perhaps more probably, Nereid could be a captured body — a large asteroid or small escaped satellite — that underwent a collision in the vicinity of Neptune, lost energy and was captured by the planet into its present highly-eccentric orbit.

Now I am going to deal with the origin of dwarf planets, the bodies that Simon introduced last week. We have already identified the Moon, Mars, Mercury and Triton as onetime satellites of the colliding planets but there would have been many more. There are now six bodies classified as dwarf planets — Ceres, Pluto, Eris, Makemake, Haumea and, a recent discovery, V774104. They all have masses within the range of solar-system regular satellites and we identify them as ex-satellites of the colliding planets.

The orbits of bodies of satellite mass are modified by the Type 1 migration process in which the body maintains

---

[5]McCord, T.B. (1955) *Astronomical Journal,* **71B**, 585–590.

contact with the medium, but the effectiveness of the process depends on the mass of the satellite as well as the density of the medium. Since orbital evolution is slow for smaller-mass bodies it is unlikely that orbital evolution will progress to the round-off stage within the lifetime of the circumstellar disk. When I spoke about the round-off and decay of the original capture-theory planetary orbits I showed you how the semi-major axis and eccentricity varied with time but what I did not show, which is in this figure (Figure 10.8), is the way that the perihelion varies with time. The perihelion increases up to the time that round-off occurs and thereafter falls as the circular orbit decays. Satellites with orbits that evolved to the extent that the perihelion was within the Kuiper Belt would survive while others would eventually be swept up by a major planet or expelled from the Solar System. The exception is Ceres that was left in an orbit within the asteroid belt.

There could have been many other ex-satellites, possibly some more massive than the present dwarf planets. Many of these might have escaped from the Solar System directly from the result of the collision and others have been

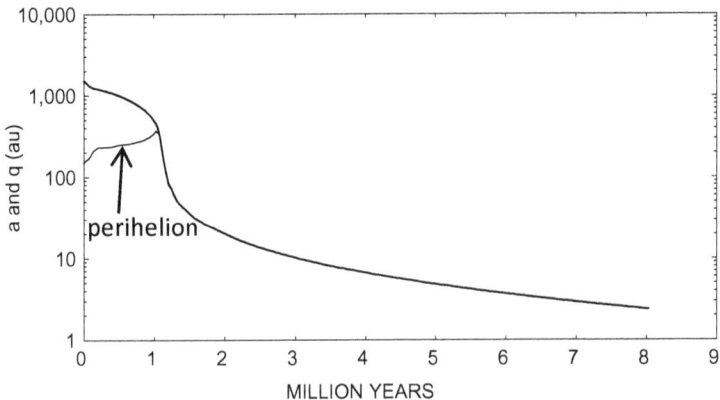

Figure 10.8   Planetary decay, showing the increase in perihelion up to the time of round-off

absorbed by major planets. Others, yet to be discovered, might be in outer regions of the Kuiper Belt. Another possibility is that one or more of them may have reached the Oort Cloud; it has sometimes been suggested that the tendency of some new comets to come from similar directions with similar orbital parameters may be due to the presence of major perturbing bodies within the Oort Cloud itself.

If dwarf planets were redefined as being ex-satellites of the originally-formed planets, large enough for their self-gravity to mould them into hydrostatic equilibrium and in heliocentric orbits then, according to the model suggested here, at present there would be eight — Ceres, Pluto, Eris, Makemake, Haumea, V774104, Mars and Mercury.

There is just one more topic I would like to mention today and that is the reason for the LHB. Simon told us that the Nice Model explained it as due to the ice giants being propelled into the planetesimal region outside the initial orbit of Uranus, but the timing seems all wrong. It seems likely that any plausible model of solar-system formation would have had it in a more-or-less settled condition 500 million years after its formation. However, evidence for the LHB is incontrovertible so we must consider how it might have happened.

The most likely cause, in my opinion, is that it was due to an inward rain of Oort-cloud comets due to massive perturbation by some body external to the Solar System. There are two possible candidates for the perturbing body, a star or a giant molecular cloud (GMC), and the latter, which would have had a far larger effect, is the more likely. GMCs are very variable in their characteristics but a typical one would have 10^5 solar-mass and a diameter about 50 pc. It has been estimated that the Solar System might have passed through up to four GMCs during its

lifetime.[6] Their perturbing effectiveness is due to the fact
that they have a clumpy structure so that as the Solar
System moves through a GMC it is subjected to periods of
large perturbation as clumps pass by. A passage through
a large GMC 500 million years after its formation is a
likely event which would have the effect of decimating
the total population of Oort-cloud comets and producing
the LHB. With the comet population greatly depleted,
any subsequent passage through a GMC would give
bombardment on a much smaller scale. There is evidence
from lunar cratering of lesser episodes of bombardment
2.0, 0.8 and 0.5 billion years ago, all much less intense than
the LHB and each slightly less intense than the previous
one.[7] Although other possible causes of the LHB cannot
be discounted the scenario given here is plausible and fits
the observations.

Solomon: Have you concluded your account of the effects of the
planetary collision? You seem to have covered everything
as far as I can tell.

Steven: No, not quite. I still have one more thing to talk about
but it is too big a topic to start now.

Solomon: Simon, Steve was suggesting that you should explain the
composition of meteorites in terms of the Nebula Theory.
Will it take long?

Simon: No, it is quite straightforward. Differentiated asteroids
are formed by the collision and fragmentation of *parent
bodies*, large collections of planetesimals several hundred
kilometres in diameter. Because they would contain short-
lived radioactive material, such as aluminium-26, they
would be molten and, since they are big enough to have

---

[6]Clube, S.V.M. and Napier, W.M. (1983) *Monthly Notices of the Royal Astronomical Society*, **208**, 575–588.

[7]Fagan, A.L. *et al.* (2015) *Unravelling the Bombardment History of the Earth–Moon system — 2 billion years ago*. 48th Lunar and Planetary Science Conference.

reasonably large internal gravitational fields, material with them would be differentiated by density.

Steven: That is ad-hoc if you like. Take planetesimals of asteroid size, put them together to form larger bodies and then smash them up to get bodies of the original size. How many parent bodies are there and what is the chance of them colliding and when did they collide?

Solomon: Let's not get too tied up with this topic. Despite what you say, Steve, what Simon has described is possible. You have some more to tell us so that is our topic for next week. Do you know what it will be about, Simon?

Simon: No. As far as I can tell we have both covered the field.

Solomon: Righto Steve, you have us guessing but that will add to the anticipation of next week's topic.

The three friends chat while finishing their ale and then leave.

The Eleventh Meeting

Steven Describes the Origin of Isotopic Anomalies in Meteorites

On 15ᵗʰ April 2016, Steven and Solomon are sitting in their cubicle chatting about the previous week. Simon arrives with three full glasses on a tray and then sits beside Solomon facing Steven.

Solomon: Right, Steve, tell us what you're going to talk about this evening. I've spent the last week wondering what it might be.

Steven: I'm going to tell you about isotopic anomalies in meteorites, what they are and how it is proposed that they came about. In my opinion they are not given the prominence that they ought to have as a clue to the processes that went on in forming the Solar System and about events in its early history. By-and-large it is ignored by the nebula-theory community and meteoriticists — people who study meteorites — certainly don't relate it to events that formed the Solar System but treat it as something that happened within, or to, the already-formed Solar System.

As it happens I gave a talk on this topic on Tuesday to a local astronomy society and I first had to explain to them the basic structure of an atom in terms of a nucleus containing protons and neutrons and the accompanying electrons and then what an isotope was before I could begin. I used carbon as my example and showed them this figure (Figure 11.1). I explained to them that their

177

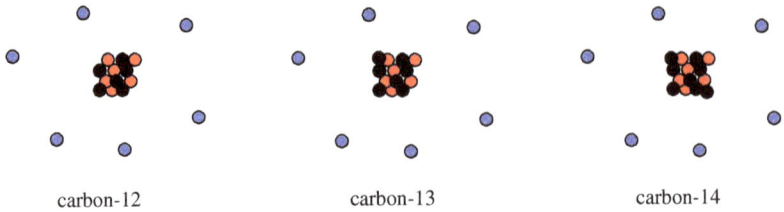

Figure 11.1   The structures of carbon-12, carbon-13 and carbon-14. Red = proton, black = neutron, blue = electron

bodies were chock-a-block full of carbon-12 and carbon-13 atoms in the ratio 89.9:1 and there was also a tiny amount of carbon-14, which was radioactive and transformed into nitrogen-14 with a half-life of 5,730 years. I think that I managed to put over the idea that the chemical nature of an atom was determined by the number of protons in the nucleus but that there could be stable isotopes with different numbers of neutrons. Anyway, I don't have to tell you professional physicists these things but I always enjoy the challenge of explaining scientific ideas to people with just a little background in science.

Simon: I know what you mean Steve. I also give talks to local astronomical societies and I find that looking for ways to put over the science simply, but without distorting the science to the point where it becomes untrue, can clarify my own understanding.

Steven: That is exactly my experience, Simon. We know that meteorites are derived from asteroids; we can determine the reflection spectra from asteroids illuminated by the Sun and compare them with reflection spectra from meteorites measured in the laboratory. Very good matches can be made, which tell us something about the composition of asteroids. In the planetary-collision interpretation of asteroids, they are fragments of pre-existing planets. This contrasts with the nebula-theory idea that planets are formed from planetesimals, which are essentially similar to asteroids.

Now, while there is no reason to believe that all asteroids are the same chemically — chemistry depends on many factors such as temperature, pressure and local concentrations of elements, all of which can be variable from body to body — we might expect them to be very similar in terms of isotopic composition. Isotopic composition can be very slightly changed by diffusion processes that depend on the different masses of isotopes, but only very small changes can be made in this way. If we find a meteorite with the ratio of C-12:C-13 very different from 89.9:1 then we might wonder what had happened to it or, indeed, whether it was really from the same original source as terrestrial material.

It is found that for many elements the isotopic compositions in some kinds of meteorites are very different from those of terrestrial material; where this occurs for a particular element we describe the difference as an *isotopic anomaly*. I will now describe a number of isotopic anomalies that have been found for some lower-mass elements.

In some chondritic meteorites — those containing small glassy spheres called chondrules — there are grains of the mineral silicon carbide, SiC. When the C-12:C-13 ratio is measured in grains of SiC from different meteorites large variations are found. Ratios are found that are substantially less than 89.9:1 — down to about 20:1 — and since these samples have a greater proportion of C-13, the heavier isotope, this isotopic anomaly is referred to as *heavy carbon*.

The standard explanation by meteoriticists for this anomaly is that it is due to an influx of grains into the Solar System from distant carbon stars. Carbon stars are known from their spectra to contain heavy carbon, with the heaviness differing from one carbon star to another, and it has been suggested that the various heavy-carbon isotopic anomalies in silicon carbide grains can be explained if they originated in six or more different carbon stars. These grains are assumed to have drifted through interstellar space, entered the Solar System and then somehow became incorporated in asteroids.

Silicon carbide grains also contain nitrogen trapped in interstices in the grains. The ratio of the two stable nitrogen isotopes in the terrestrial atmosphere, N-14:N-15, is 270:1. Most of the nitrogen from silicon carbide is *light nitrogen* with a ratio of up to 2,000:1 but, curiously, *heavy nitrogen* also occurs rarely with a ratio as low as 50:1.

A very intriguing and important isotopic anomaly is that for oxygen. There are three stable isotopes with ratios in terrestrial samples, such as water, the atmosphere and silicates, given by O-16:O-17:O-18 = 0.953:0.007:0.040. This mixture is known as Standard Mean Ocean Water (SMOW). There are very small variations away from SMOW in terrestrial samples but these are due to physical conditions — for example, the existence of high temperature gradients in some materials — and the reason for them is well understood. Oxygen from Moon-rocks seems compatible with that on Earth.

There are two kinds of meteorites that contain oxygen isotopic ratios that cannot be explained by processing terrestrial oxygen in some way. These meteorites are ordinary chondrites and carbonaceous chondrites, the latter being stony meteorites that contain a great deal of volatile material including water, although not in liquid form but as water of crystallization of minerals, one example being the mineral serpentine. The oxygen anomalies in these meteorites can be explained as the addition of different amounts of pure, or nearly pure, O-16 to SMOW. The problem is to find a source of pure O-16.

The usual explanation again involves grains drifting across interstellar space and entering the Solar System. The O-16 is assumed to have been produced by the action of alpha particles (the nuclei of helium atoms, He-4) on

C-12 in distant stars. This pure O-16 is then incorporated into dust grains that travel to the Solar System and become incorporated in asteroids. Subsequently, normal solar-system oxygen infiltrated the grains displacing most, but not all, of the O-16 so giving a final content that has a surplus of that oxygen isotope.

As for oxygen, there are three stable isotopes of magnesium in the terrestrial ratios Mg-24:Mg-25:Mg-26 = 0.79:0.10:0.11. Carbonaceous chondrites, which are generally very dark in colour, contain white inclusions of high-melting-point minerals known as Calcium Aluminium-rich Inclusions (CAIs) and the magnesium in these is anomalous in that in each grain of a particular CAI there is an excess of Mg-26 that is proportional to the amount of aluminium in the grain.[1]

There is only one stable isotope of aluminium, Al-27, but, as Simon previously mentioned in relation to the heating of the satellite Phoebe, there is the radioactive isotope of aluminium, Al-26, with a half-life of 717,000 years, which decays to give Mg-26. The aluminium in the material from which a particular CAI originally formed contained a small component of Al-26, in the general range of 1 part in 100,000 to one part in 10,000. The Al-26 then decayed leaving behind Mg-26 that gave an excess of Mg-26 when combined with the normal magnesium that was present. Different mineral grains in a particular CAI would have different chemical compositions but the aluminium in them will all be derived from the same source and hence have the same ratio of Al-26 to Al-27. The excess of Mg-26 in a particular mineral grain within the CAI would then be proportional to the amount of aluminium (hence also of Al-26) that it contained.

[1]Lee, T., Papanastassiou, D.A. and Wasserburg, G.T. (1976) *Geophysical Research Letters*, **5**, 109–112.

It has been suggested that Al-26 was produced in a supernova nuclear explosion that was responsible for triggering the formation of a cool dense star-forming cloud, within which the Sun and other stars were formed. However, it takes about 200 million years for the cloud to form, by which time the Al-26 would have disappeared. For that reason it has been proposed by meteoriticists that, by chance, just before the Solar System formed, there was a second supernova somewhere in its vicinity that injected Al-26 into the Solar System.

For our final element in this description of isotopic anomalies, I take neon, which forms a small part of the terrestrial atmosphere. There are three stable isotopes with terrestrial ratios Ne-20:Ne-21:Ne-22 = 0.905:0.003:0.092. Since neon is an inert gas it does not produce chemical compounds, so when it occurs in meteorites it is in the form of individual neon atoms trapped in atomic-size cavities, but they can be released by heating the meteorite, which gives the neon atoms enough energy to percolate through the grain and escape. For this reason, if neon is found in meteorites then it is certain that these meteorites were not substantially heated for any length of time after the neon was incorporated — otherwise the neon would have escaped. The same argument can be used for other gases trapped in the interstices of meteorites.

The isotopic compositions of neon from different solar-system sources are very variable and it has been deduced that there may be three separate sources with different compositions and the neon we normally measure in meteorites comes from different admixtures of these sources.

There is one neon anomaly that cannot be explained as an admixture from three basic sources. Some meteorites contain pure, or almost pure, Ne-22, which is a 9.2 percent component of terrestrial neon. This anomalous neon is called neon-E. It is not possible for Ne-22 to be separated out from a mixture of isotopes so some other explanation for its presence is required. Sodium has only one stable isotope, Na-23, but there is a radioactive isotope of sodium, Na-22, which decays into Ne-22. One suggested scenario is that Na-22 was produced in a supernova, possibly the same one that gave Al-26, and then was incorporated, together with stable sodium, in minerals. The Na-22 decayed and the resultant Ne-22 was trapped in interstices within the mineral grain. However, there is a fundamental problem with that scenario; Na-22 has a very short half-life — 2.6 years. This means that the radioactive sodium has to be produced in a supernova and then incorporated into a *cool* solid body within a period of 10–20 years. If the body were not cool shortly after the incorporation of Na-22 then the decay product, Ne-22, would not have been retained.

There are numbers of other isotopic anomalies in meteorites but I will concentrate in explaining these I have given.

Solomon: You have given us the observational background for your explanation of these isotopic anomalies so now seems to be a good time to take a break. I've been taking notes but I would like to sort them out. Is this stuff new to you Simon?

Simon: Not really, but I have never explored the topic in any depth since it does not impinge directly on the process of planet formation or how the bodies of the Solar System are arranged. I'll be interested to see how Steve manages to connect this topic to planet formation in a logical way.

The three sit quietly drinking for several minutes then, when the glasses are all empty, Solomon goes to the bar and brings back three full glasses.

Solomon: Simon seems to think that isotopic anomalies have no relevance to solar-system formation or how the system evolved. Can you convince us otherwise, Steve?

Steven: The explanation for the isotopic anomalies I have described is yet another outcome of the planetary collision[2] and hence is very relevant to solar-system formation. The deuterium-rich layer in the colliding planets was taken as having a D/H ratio of 0.01, although, consistent with observations, it may have been greater. Once the temperature in that layer reached 3×10^6 K then D–D reactions took place in an explosive fashion and the temperature quickly reached 10^8 K, at which stage other reactions involving heavier elements could take place. By the time the temperature reached somewhere around 5×10^8 K, but perhaps higher depending on the D/H ratio, the deuterium was exhausted but the temperature continued to rise although at a slower rate.

In 1995, Holden and Woolfson considered the outcome of such an explosion in a mixture of silicates and deuterium-rich material in terms of the isotopes that would be produced.[3] Their computer program incorporated the reaction rates, given by Fowler, Caughlan and Zimmerman,[4] for 548 different nuclear reactions, involving the most common elements that would be present in a silicate–ice mixture — up to and including iron in the periodic table. The effect of radioactive decays had to be included since some of the products of the reactions are radioactive and what is now observed includes the end-products of their decay. At the time of the Holden and Woolfson work the high D/H ratios in the grains of star-forming clouds and protostars was not known so they took the D/H ratio as 0.016, the value for Venus. In 2011 Woolfson repeated the calculation with

---

[2]Woolfson, M.M. (2013) *Earth, Moon and Planets*, **111**, 1–14.
[3]Holden, P. and Woolfson, M.M. (1995) *Earth, Moon and Planets*, **69**, 201–236.
[4]Fowler, W.A., Caughlan, G.B. and Zimmerman, B.A. (1967) *Annual Review of Astronomy and Astrophysics*, **6**, 525–570; (1975) *Annual Review of Astronomy and Astrophysics*, **13**, 69–112.

D/H = 0.01 and obtained different, but similar, results that are the ones I report here.[5] By the way, some time ago Simon referred to observations sometimes confirming previous assumptions — for the Nebula Theory the existence of circumstellar disks. Holden and Woolfson assumed a high D/H ratio in ices before observations confirmed that it was so.

The explosion produced large quantities of C-13 and also the radioactive isotope of nitrogen, N-13, which decays to C-13 with a half-life of 9.97 minutes. The full range of heavy-carbon observations can be explained in terms of this production of C-13. This figure (Figure 11.2) shows the concentrations of C-12 and C-13 during the explosion as a function of the temperature; the C-13 concentration includes the contribution of N-13. It will be seen that when the temperature exceeds 3×10^8 K there is a sharp increase in the amount of C-13 and a smaller reduction in C-12.

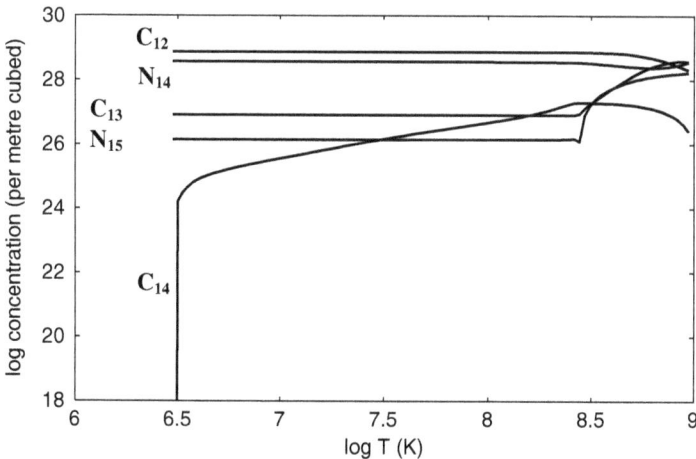

Figure 11.2 The concentrations of isotopes of carbon and nitrogen as the temperature within the nuclear-reaction region varies

[5]Woolfson, M.M. (2011) *On the Origin of Planets: By Means of Natural Simple Processes* (Imperial College Press, London).

The ice in star-forming clouds contain considerable quantities of nitrogen in the form of ammonia, NH_3. The figure shows that there is a small reduction in the amount of N-14 as the explosion progresses, although the amount picks up again at very high temperatures. The concentration of N-14 includes the contribution of radioactive O-14 that decays to N-14 with a half-life of 70.6 seconds. Over a long period, the N-14 is augmented by the decay of C-14, the isotope used for carbon dating in archaeology, with a half-life of 5,739 years. To explain light nitrogen from silicon carbide grains, we start with some N-14 and N-15 trapped inside the grains once they were cool enough to retain the gas. These would not have corresponded to light nitrogen with the high N-14:N-15 ratios that are observed but over the course of the next few tens of thousands of years C-14, which had been part of the silicon carbide that had formed, will have decayed with the resultant N-14 boosting the amount of that isotope, so giving light nitrogen. This would be particularly effective if the original amount of the two stable isotopes had not been very great for then the C-14 contribution can make a large proportional change in N-14.

Towards the end of the explosion, there is a much greater than ten-fold increase in the amount of N-15 present. This will be produced by nuclear reactions involving the isotopes of elements, which may not be uniformly distributed within the explosion region so N-15-rich pockets can form. Silicon-carbide becoming cool within one of these pockets could take up nitrogen with a large component of N-15 and even after the decay of C-14 the result can be the occasional occurrence of heavy nitrogen.

I now show you how the three stable isotopes of oxygen vary during the explosion (Figure 11.3); the O-17 and O-18 concentrations include the contributions of two short-lived fluorine radioactive isotopes — F-17 and F-18 — that decay to O-17 and O-18, respectively. It will be seen

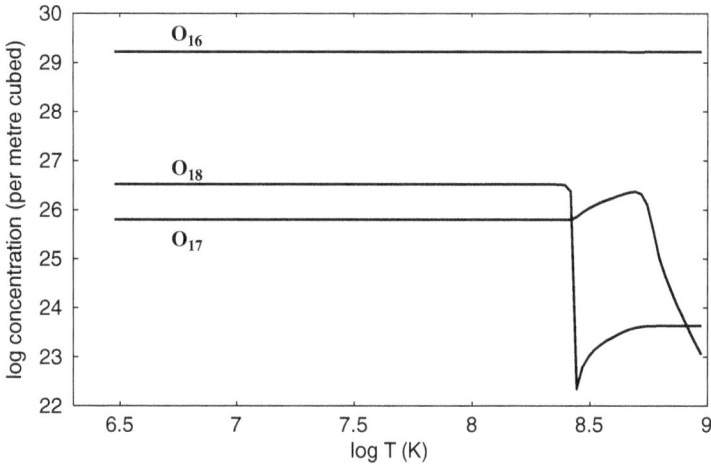

Figure 11.3 The variation of the isotopes of oxygen (including radioactive fluorine) with temperature

that at high temperatures, above about 6×10^8 K, the concentrations of O-17 and O-18 are greatly diminished, leaving virtually pure O-16. This oxygen, mixed with normal SMOW in various proportions explains the observed oxygen isotopic anomaly.

The Mg-26 anomaly, i.e. an excess of that isotope proportional to the amount of aluminium present, depended on the presence of radioactive Al-26. This figure (Figure 11.4) shows that significant production of Al-26 after the temperature reaches 4×10^8 K. However, the situation is rather confused since the concentrations of Mg-25 and Mg-26, which include the contributions of various radioactive decays, also vary greatly at the end of the temperature range. The observations seem equally confused; Wasserburg, Lee and Papanastassiou, who reported the Mg-26 excess anomaly, also report situations where there is a *deficit* of Mg-26,[6] which ties in with the sharp

---

[6]Wasserburg, G.T., Lee, T. and Papanastassiou, D.A. (1977) *Geophysical Research Letter*, **4**, 299–302.

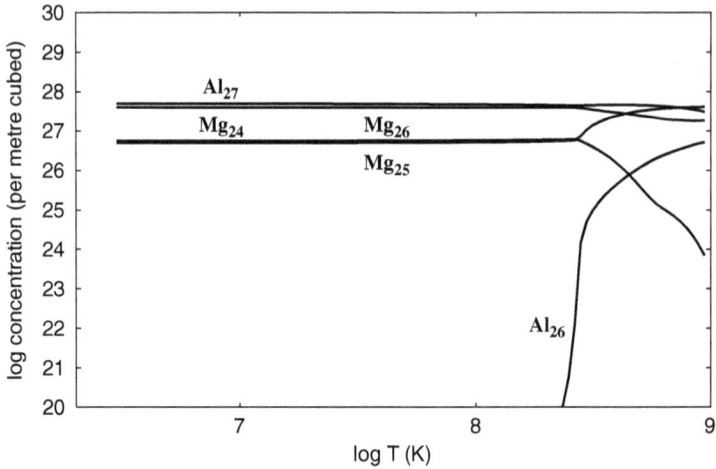

Figure 11.4   The production of magnesium and aluminium isotopes during the nuclear explosion

fall-off of that isotope at higher temperatures. The form of magnesium anomaly probably varies with location in the explosion and local concentrations of isotopes of other elements, reactions involving which produce the stable isotopes of magnesium.

A sufficient quantity of Na-22 was produced in the explosion to explain the production of neon-E. We noted that it was suggested that Na-22 was produced in a supernova and was then incorporated in cold material on a timescale of 10–20 years. This scenario is somewhat more plausible for Na-22 production in a planetary collision than in a supernova. The scale of a planetary collision is small by astronomical standards and it can be shown that expansion, the formation of grains and cooling would all take place within hours or days.[7]

This completes my description of isotopic anomalies. Although, as Simon has indicated, it has nothing to do

---

[7]Woolfson, M.M. (2011) *On the Origin of Planets: By Means of Natural Simple Processes* (Imperial College Press, London).

directly with planet formation, or the arrangement of bodies in the Solar System, it is an important feature of some solar-system material. The fact that it ties in with a postulated scenario, a planetary collision, that explains so much else about the Solar System, is also significant.

Solomon: I found that interesting and convincing — what do you think, Simon?

Simon: As far as I know this explanation of isotopic anomalies has not been adopted by the meteorite community, who prefer other explanations.

Steven: Yes, they do — a series of ad-hoc explanations involving dust grains floating through interstellar space and various forms of particle bombardment to explain individual isotopes. To prefer that to an explanation that produces them all in a single event, which also explains much else, is unscientific in my opinion.

Solomon: Do either of you have anything more to add in your descriptions of how planets form, the properties of those planets and the structure of the Solar System?

Simon: No. I think that my account is as complete as I can make it without going into fine detail that would confuse rather than clarify.

Steven: The same applies to me.

Solomon: Your ideas have been presented over several weeks and there has been a fragmented presentation of both theories. I suggest that when we next meet you should give summaries of the theory you favour to give an idea of the connections between various aspects of each of them. I propose that Simon should talk first and then Steve.

Simon: Yes, that is a good idea. I'll try to give a coherent picture of the Nebula Theory and explain how the various aspects of it link together.

Steven: And I'll do the same for the Capture Theory.

They chat a while and then leave.

The Twelfth Meeting

Simon and Steven Summarize the Two Theories

On 22nd April 2016, Simon and Solomon are sitting in their cubicle. Simon is sorting out material for his presentation. Steven approaches from the bar with three full glasses.

Steven: The landlord has just taken delivery of a new-on-the-market real ale called *York Passion* and he has sent these with his compliments with a request that we give our opinion of it.

The three sample the ale for some minutes.

Steven: I think that this is the best I've ever tasted. It leaves a warm lingering afterglow that I find rather pleasant.

Simon: Yes, it does. I'll make a note of the name and suggest it to my neighbouring watering-hole.

Solomon: It *is* good. We'll tell the landlord in due course but first let's get down to business. Simon, summarize the Nebula Theory for us.

Simon: I'll start by briefly reminding you of the history of ideas about the origin of the Solar System, something I have agreed with Steve and he has given me some of his illustrations. Up to 1995, when the first exoplanet was discovered around a Sun-like star, the aim in this field was to explain the origin of the Solar System. Laplace had proposed his nebula-based theory, illustrated in this figure (Figure 1.2) and since he was describing how a star inevitably formed with a surrounding disk it was implicit that what happened in forming the Sun and planets would also occur with other stars. There are really two types of theory that can be envisaged. The first, like Laplace's theory, would make planetary systems common. However, even with many planetary systems being formed the Solar System might still be special in the sense that one of the planets, our Earth, could support life and that we humans evolved that could consider how the Solar System began and also many other scientific problems. The second type of theory would depend on an event with such a small probability of occurring that planetary systems might be extremely rare, or even that the Solar System was unique in the galaxy, if not in the Universe. Laplace's original nebula theory foundered in the mid-19th century because of the angular momentum problem — the Sun with 99.86 percent of the mass of the system contains only 0.5 percent of the angular momentum in its spin. No mechanism could be found or envisaged at that time that would give such a partitioning of mass and angular momentum.

In the first half of the 20th century, a number of different ideas were advanced, some of which Steve mentioned when he first described the Capture Theory. There was the Chamberlin and Moulton idea of a star passing by the Sun when large prominences were occurring, as shown here (Figure 2.1) and pulling them out further by tidal forces. The material of the prominences then condensed by stages to give the planets. The legacy of that theory is the formation of *planetesimals*, a term now used to describe the small bodies produced by fragmentation of the dust disk in the modern Nebula Theory. The Chamberlin and Moulton theory was untenable for a number of reasons, the main one being that it was then thought that spiral nebulae were stars within our galaxy rather than, as we now know, other galaxies.

Following on from Chamberlin and Mouton, and somewhat related to it, there was Jeans' tidal theory, again described by Steve and shown in this figure (Figure 2.3), which was also abandoned because of basic faults; it was shown that it could not produce planets far enough from the Sun and also that the material pulled out of the Sun would be too hot to condense into planets. It also would have made planetary systems quite rare, something we now know not to be true.

An important new concept for nebula development was proposed by Hoyle in 1960 with the idea that magnetic linkage in a gap between a collapsing central core and a disk, as seen here (Figure 3.1) could transfer angular momentum outwards. While the original Hoyle model had some problems, it introduced a new mechanism, of which Laplace and mid-19th century scientists would have been unaware, that, in a developed form, could possibly make nebula ideas viable in terms of angular-momentum distribution. Another, and quite effective, way of transferring angular momentum was the mechanical process described by Lynden-Bell and Pringle. In a system where energy was being lost but angular momentum had to remain constant, material close to the spin axis would move inwards while that further out would move outwards, tantamount to a transfer of angular momentum from inner to outer material.

A planetary-science development in the 1960s was the realization that many characteristics of meteorites could be understood in terms of condensation sequences, i.e. minerals condensing out of a cooling hot vapour in the order of descending melting points. Initially this raised the image of a hot nebula surrounding a newly-forming star, although later it was realized that planets could not form from hot material. Nevertheless, planets could form once the nebula had cooled. Anyway in the 1960s development began on what was initially referred to as the Solar Nebula Theory but later, when exoplanets were discovered, as the Nebula Theory.

Sometimes the development of a theory does not follow a logical temporal path in the sense that an assumption is made in developing a theory and only later is the assumption justified by observation. So, it was with the initial development of the Nebula Theory that assumed that there would be a disk surrounding a young star, but it was not until the late 1970s that such disks were actually detected. Naturally, the observational confirmation of something that had previously just been conjecture gives a tremendous boost to the theory being developed. Later it was found that the lifetime of the gaseous disks was quite short — up to 10 million years but more typically 3 million years. This constrained the amount of time for a giant-planet core to form since, after it formed, there had to be gas present for a massive atmosphere to be captured. However, after the gas has dispersed there is left behind a debris disk, consisting of planetesimals, that can be extremely long-lasting and, as it turns out, this debris disk is important for the evolution of the planetary orbits. Since there is now no doubt that young stars have surrounding disks, we can take this as our starting point for considering the Nebula Theory. The problem of how the distribution of angular momentum between star and disk occurred may be an interesting one but we know from observations that it must have a solution.

The first stage in forming planets from disk material is that the dust component must settle towards the mean plane of the disk under gravitational forces. Initial theoretical work on this mechanism considered 0.1 micron dust particles and the estimated settling time was many millions of years. The buffeting of the tiny dust particles by atoms and molecules of the gas gave Brownian motion in which the particles jiggled to-and-fro in a random fashion and settling was extremely slow. Two factors improve the situation. The first is that the dust particles stick together and form larger entities that settle more quickly and the second is that, although sub-micron particles are the easiest to detect because of the way they absorb and scatter light, there are larger particles present; in his description of satellite formation Steve considered particles up to 5 micron in size. These settle far more quickly and sweep up smaller particles as they move towards the mean plane. The general conclusion is that dust settling will only take a small fraction of the lifetime of the disk.

The next stage of the process is the gravitational break-up of the dust carpet to form planetesimals. When he described satellite formation Steve told us of the condition for the dust carpet to fragment, when the local dust density in a particular region reached a value high enough for a condensation there to resist disruption by tidal effects due to the central body. Applying this criterion to the planet-forming scenario, dust condensations, which we call planetesimals, form quite quickly, varying in size from hundreds of metres to many tens of kilometres.

The aggregation of these planetesimals to form terrestrial planets or the cores of major planets is the most challenging stage of the planet-formation process. For planetesimals to combine when they come together it is necessary to have an environment, with very little turbulence so that their approach speeds are small. In a turbulent environment, they would collide violently, either bouncing apart or even becoming fragmented and scattering in all directions. This need for little turbulence means that planetesimals are in near-Keplerian orbits and the aggregations that give terrestrial planets or major-planet cores would also be expected to be in near circular orbits around the central star.

As Steve has rightly pointed out, using Safronov's equation for planetary cores gives unacceptably-long formation times for the outer major planets of the Solar System if they formed in their present locations. However, although the Wetherill and Stewart theory of the runaway accretion of planetesimals, as they originally presented it, may not be completely valid, I think that some aspects of the theory may apply and that the formation of cores out to the distance of Jupiter's orbit could occur within the lifetime of the gas disk. Once the cores have formed the acquisition of a gaseous envelope to produce a major-planet atmosphere takes a comparatively short time. It is then necessary for outward migration to occur and this can happen by interaction of the newly-formed planets with both the residual gas and the planetesimal population of the disk, particularly with the latter.

Steve accepted the validity of the Nice Model simulation but challenged the starting point, so let us see how this could have come about. If we take the mass of the circumsolar disk as, one-tenth of the mass of the Sun, i.e. 2×10^{29} kg, then this would give the mass of dust as approximately 2×10^{27} kg or about 330 Earth-masses. If the majority of that went into producing planetesimals then there would be a large mass of planetesimals left over after the terrestrial planets and cores of the major planets had formed, which would take a few tens of Earth-masses at most. Now, as I previously asserted, although the Wetherill and Stewart model for runaway growth might not be as effective as they claim I still think that some aspects of it would be valid and terrestrial planets and major-planet cores could all form within, say, 5 au of the Sun within the lifetime of the disk. If so then we have the planets formed, with atmospheres for the major planets, in the inner part of the Solar System accompanied by a large mass of planetesimals. The density of the disk, and hence the density of planetesimals within it, will be highest closest to the Sun and decrease monotonically with increasing distance and they would stretch out to the 35 au or so envisaged by the Nice Model. With the areal density of planetesimals decreasing outwards, for the reasons I gave when I described the Nice Model, the planets would be propelled outward and by the time the remaining planetesimals in the inner part of the system had either been absorbed by collisions with planets, or thrown out to large distances, the starting point for the Nice Model could have been reached. The Nice model requires only a small part of the original planetesimal population, the rest having been involved in the interactions that produced the starting point for the Nice model.

You will recall that the technique for investigating the Nice Model was to set up a number of different starting points with distances from the Sun for the major planets satisfying a set of constraints. A high proportion of trials were successful in the sense that the final outcome had a passing resemblance to the present arrangement of major planets. The starting points shared the characteristic that the initial arrangement had Neptune's orbit within that of Uranus. Due to planetary interactions, after a few million years a 2:1 orbital-period resonance between Saturn and Jupiter was established. The consequent magnified perturbation of Saturn disturbed the ice giants either by directly expelling Neptune beyond Uranus by a direct Saturn–Neptune interaction or by so greatly disturbing Neptune that it directly interacted with Uranus in such a way that they more-or-less exchanged places. The action of planetesimals on the planets, particularly the ice giants, was to cause them to migrate outwards with orbital round-off due to dynamical friction.

The Nice-Model process explains a large number of features of the Solar System in terms of the final destination of individual planetesimals or aggregations of planetesimals. For example, when Uranus and Neptune are pushed out into the surrounding region of planetesimals they produce a sudden increase in the number of planetesimals entering the inner Solar System, which explains the LHB, which shows up on many bodies in the inner Solar System, particularly on the Moon. Steve has questioned the timing of this, but when it happens will depend on how long the initial migration process took to reach the Nice-Model starting point. It may take a few hundred million years but, obviously, this is something that should be explored.

During the period leading up to the start of the Nice Model many accumulations of planetesimals would have occurred giving either future dwarf planets, some of which have evolved into orbital resonance with Neptune, or parent bodies, the differentiated fragments of which formed asteroids. These latter would still function like planetesimals in their effect on the larger bodies. It is quite possible that the terrestrial planets formed during the evolution towards the starting point for the Nice Model. The formation of Mars may have been inhibited by the proximity of Jupiter, so explaining its small mass. Two other Mars-size bodies could have formed, one of which collided with the Earth, giving rise to the Moon, and the other, struck by a smaller body that stripped off part of its mantle, producing Mercury.

A small proportion of the inward-moving planetesimals or asteroids, by collisions with each other and interactions with planets, would have been deflected into the Asteroid Belt. Others would have taken up orbits as Trojan asteroids accompanying Jupiter and Neptune.

The Oort Cloud comes from the period when there were many planetesimals in the inner part of the Solar System, and there was evolution towards the Nice-Model initial state. Large numbers of planetesimals were thrown out great distances by interaction, mainly with Jupiter but perhaps also with other major planets.

On the question of regular satellite formation, Steve and I are in general agreement that they formed in a circumplanetary disk in much the same way that the Nebula Theory forms terrestrial planets and major-planet cores, Again, we agree about irregular satellites being captured bodies, usually due to collisions of asteroids in the vicinity of a planet or more rarely due to capture in a manybody environment. Our one disagreement concerns the formation of the Moon and the Earth–Moon relationship, I, like the majority of planetary scientists, accept the Benz, Slattery and Cameron model involving the oblique collision of a Mars-size body with the Earth and the formation of the Moon by the aggregation of fragments left in Earth orbit.

Steve made the question of spin-orbit misalignment a big issue and suggested that it was a severe challenge to the Nebula Theory. Equally, he has emphasized in his presentations the idea of planetary interactions, even to the extent of having two of them collide, and has also mentioned the importance of the dense embedded state of a star-forming region in which stars can approach each other closely. In such circumstances, it would be remarkable if all planets remained in their original state of having close-to-zero spin-orbit misalignment. I do not see the observations as presenting any great challenge to the Nebula Theory.

On the frequency of planetary systems, it seems that the formation of a circumstellar disk is normal and hence one would expect planet formation to occur via the nebula-theory process if the disk has suitable physical characteristics — for example, total mass and sufficiently-high areal density. It is not possible to produce a numerical prediction of the proportion of stars with planets, but that planet formation should be common is almost certain.

This completes my general description of planet formation and solar-system evolution according to the Nebula

Theory. It is impossible to dot every *i* and cross every *t* in describing the formation of planets since it is bound to involve a number of complex interacting mechanisms. However, the general pattern of planet formation, linked to the presence of circumstellar disks, which is observationally confirmed, seems to me to be soundly based and has wide acceptance in the astronomical community.

Solomon: What do you say to that, Steve?

Steven: Ptolemy's theory of an Earth-centred Solar System had wide acceptance in the astronomical community — but it was wrong! You tend to get bandwagon effects in science where, once an idea becomes established, more and more people follow it until eventually it achieves the status of an unchallengeable truth. Anyone who believes otherwise is regarded as eccentric and any alternative model he, or she, may propose is never considered objectively and may be virtually ignored. The lack of objectivity is not helped by the fact that, once a theory has been developing for a considerable period, people will have made a great investment in time and effort in developing it and then there is a natural tendency to protect the theory by either unfairly criticizing or ignoring any threatening alternative.

Solomon: Well, you will now have a chance to express your ideas and I can promise you that, for my part, you will not be ignored. But first let us have a few quiet minutes while we empty our glasses. I think you'll agree with me that York Passion is really good stuff and we should tell the landlord so.

Simon: Yes, it really is good.

Steven: I agree. This lot was a present from the landlord so I'll get the refills and tell him how good it is.

They sit silently drinking while Steven sorts out the notes for his presentation. When the glasses are empty Steven goes to the bar and returns with full glasses.

Solomon: Off you go, Steve.

Steven: Any account of the Capture Theory, and what follows from it, must start with star formation because this is what provides the scenario within which the capture-theory mechanism takes place. A supernova both compresses the local interstellar medium by the shock waves it generates and also injects vaporized material into it that condenses into the form of a fine dust. Cooling occurs both by radiation from the dust[1] and processes due to the excitation of molecules, atoms and ions by electron collision.[2] When an atomic system is excited it goes into a state of higher energy; for an ion or atom this can mean that one of its electrons is pushed into a higher-energy state or for a molecule that it distorts into a higher-energy configuration. When the system reverts into its lower energy state it emits a photon that, in a diffuse environment, escapes from the system. This loss of energy leads to cooling, a lowering of pressure and an influx of external material into the region, thus increasing its density. The cooling rate increases with higher density and decreases with lower temperature. The final outcome is a cool dense cloud in near-pressure equilibrium with the much less dense but much hotter surrounding interstellar medium.[3]

The cloud will begin a gravitational free-fall collapse, with some of the energy released by the collapse generating turbulence within the cloud. Turbulent streams of gas can collide head on, producing dense regions that become hot by compression but, since cooling is a much faster process than re-expansion, these compressed regions quickly cool and, under suitable circumstances, the dense cool region

---

[1] Hayashi, C. (1966) *Annual Review of Astronomy Astrophysics*, **4**, 171–192.

[2] Seaton, M.J. (1955) *Annual Review of Astronomy and Astrophysics*, **18**, 188–205.

[3] Golanski, Y. and Woolfson, M.M. (2001) *Monthly Notices of the Royal Astronomical Society*, **320**, 1–11.

may form a protostar. The protostars evolve, first into young stellar objects and then into main-sequence stars and new protostars are produced at an ever-increasing rate.[4] The stars fall inwards as the cloud collapses and an environment is created where condensed stars and protostars co-exist and where the stellar number density can become very high — $2 \times 10^4 \, \text{pc}^{-3}$ or even much greater. With newly formed protostars of radius 2,000 au, with free-fall times of more than 20,000 years and moving at about $1 \, \text{km} \, \text{s}^{-1}$ within a stellar environment where the separation between stars was 7–8,000 au, tidal interactions between diffuse protostars and compact young stellar objects or main-sequence stars would happen frequently. This high-density embedded state will last for a few million years until the most massive stars explode as supernovae, driving out the gas and causing the stellar cluster to expand.

Here, I remind you of what a simulated capture-theory event looks like (Figure 2.7). Planetary condensations form in the filament produced by the stretching of the protostar, some of which are captured by the star while others escape to become free-floating planets. Another way in which planets can be produced by the Capture Theory is if two turbulent streams of gas collide in the vicinity of a star, as seen here (Figure 2.10).

The initial orbits of the protoplanets produced by the capture-theory mechanism are extended and very eccentric, typically with semi-major axis 1,500 au and eccentricity 0.9. However, not all the protostar material goes into forming the filament that produces planets. Some goes into forming a disk around the star, which can take two main forms. All disks have to splay out perpendicular to their mean plane to be stable but the areal density

---

[4]Woolfson, M.M. (1979) *Philosophical Transactions of the Royal Society London,* **A 291**, 219–252.

can either fall off monotonically from the centre or take on a doughnut-like form as seen here (Figure 4.7). In most cases, the orbits round-off and decay in a few million years, with round-off being the faster process. It is found that, for a given resisting medium, both the rate and extent of the orbital evolution increase with the mass of the planet.

The decay can take a planet very close to its star but there is a mechanism that can prevent the planet plunging into the star, which depends on the star's spin period being less than the orbital period of the planet. At the other extreme, exoplanets are sometimes found at large distances from stars. Some of these have been imaged directly with the greatest distance recorded so far being 650 au. This can come about if the resisting medium is either or both of being very diffuse and very short-lived.

There are some exoplanets with large eccentricities — up to 0.97 — and these can be explained by particular prevailing conditions during orbital evolution. This can occur with a very active star where the stellar wind, pushing outwards on the resisting medium, effectively cancels out a large part of the gravitational attraction of the star so that the medium is slowed down in its orbital motion. In addition, if the medium has a doughnut form then once the orbit has evolved so that its aphelion is within the peak region the eccentricity of the orbit will steadily increase as the orbit decays,

A final aspect of orbital decay is that pairs of planets can develop commensurate orbits such as occurs for Saturn and Jupiter where the ratio of their orbital periods is close to 5:2 and Neptune and Uranus for which the ratio is close to 2:1. This is because the evolution of a planet's orbit is not only affected by the resisting medium but also by other planets. If neighbouring planets achieve commensurability of their orbits then, while both orbits continue to decay, they do so in a way that preserves the ratio of the periods.

Planets are observed orbiting one or both the stars of a binary system. The filaments within which planets form are, in general, hundreds of au from the stars so that, in the process of fragmentation of the filaments, a close binary will have a gravitational effect similar to that of a single body with slightly changing magnitude and direction. The resisting medium will encompass both stars and the round-off and decay of the orbit will resemble that round a single star until the planet approaches the binary. The final outcome will normally be an orbit around one or both of the stars, although the planet could be ejected from the binary system to become free-floating.

A very important observation for the Capture Theory is that some exoplanets have a spin-orbit misalignment — i.e. the angle between the spin axis of the star and the normal to the planetary orbit — that corresponds to a retrograde orbit. This figure (Figure 6.5) shows that all possible values of SOM occur although there is a strong bias towards small values. For the Capture Theory, there is no relationship between the spin axis of the star and the plane of the star-protostar orbit, which defines the orbital plane of the planets. For this reason alone, all possible values of SOM are possible. The bias towards small values is because some protostar material is absorbed by the star, which pulls the stellar spin axis towards the orbital rotation axis. Very little protostar material, of order of a Jupiter mass, needs to be absorbed to make a big change in the stellar spin axis, hence the heavy bias towards small values. The importance of the SOM observations is that if they had all turned out to be very close to zero then the Capture Theory would have been untenable without making extreme assumptions about the amount of protostar material absorbed by the star.

Now I want to move on to an estimate of what proportion of main-sequence stars are expected to have accompanying planets. I gave an account of a model in which a collapsing protostar moves in an environment of stars with a fixed stellar number density but all with different masses. A current estimate from observation of the proportion of stars with planets is 34 percent and the estimated proportions, for different initial protostar radii and different stellar number densities is shown in this table (Table 5.1). Considering the rather conservative conditions imposed on these simulations, taking low stellar number densities and protostar radii, and the exclusion of interactions involving colliding gas streams, it appears that the current observational estimate can easily be accommodated.

I now come to satellite formation, about which Simon and I basically agree. The nebula-theory adherents assume that in the acquisition of gas by a major-planet core a circumplanetary disk will form although, to me, the reason for thinking so is not obvious. In the capture-theory simulation, one can see that such a disk does form (Figure 2.9). An important consideration is that the capture-theory model gives a different mode of formation of planets and satellites but it has always been taken as axiomatic by most planetary scientists that the two modes of formation should be similar, albeit on a different scale. In 1919, James Jeans expressed the view that "Any theory that proposed a mechanism for producing satellites that differed from that for producing planets would be condemned by its own artificiality". Later, in 1978, the Swedish astrophysicist and Nobel Laureate, Hannes Alfvén, stated that "We should not try to make a theory for the origin of planets around the Sun *but a general theory of the formation of secondary bodies around a central body*. This theory should be applied both to the formation of satellites and the formation of planets".

To counter that view — one expressed by many others as well as by these two distinguished scientists — I showed you this table (Table 8.1) that reveals in stark terms a fundamental difference in the way that angular momentum is partitioned in the two types of system. In my opinion, although it could not be said that there *must* be two different modes of formation, at least it indicates that having two different modes of formation is perfectly reasonable. From the disk mass and size as deduced from this figure (Figure 2.9), the areal density of the disk can be estimated. Now some time ago, in discussing the Nebula Theory, I gave you a formula for the time of formation of a planetary core of mass m and density ρ from accumulations of planetesimals (Steven writes)

$$t = C \frac{P}{\sigma} \left(m \rho^2 \right)^{1/3}$$

in which C is a constant, about 0.03, P and σ are, respectively, the orbital period and areal density of the solids in the disk at the position of core formation. For forming a satellite the areal density is much higher than in a circumstellar disk and the orbital period and mass are much smaller. The density ρ may also be slightly smaller. The time of formation of satellites turns out to be of order 10^5 years, well within the expected lifetime of a circumplanetary disk so that, unlike in the case of planetary formation in this way, there are no timescale problems. Likewise, the other stages in the formation of a satellite, the settling of dust and the gravitational instability of a dust carpet to produce satellitesimals also give no difficulties.

What does, at first, seem to present a difficulty is the formation of terrestrial planets; the collapse of the protoplanetary blobs in the stretched-protostar filament depends on the mass of gas in each blob satisfying the Jeans mass criterion so that only major planets should form. The terrestrial planets, and many other features of the Solar System, can be neatly explained by the hypothesis that the early Solar System contained six major planets, the existing four plus two others that collided and provided the ingredients for most of the major features of the Solar System we see today.

The simulation of the collision of the planets, Enyo with 1.9 Earth masses and Bellona with 2.5 Earth masses is shown here (Figure 7.1). Because of the phenomenon of grain-surface chemistry operating in star-forming clouds, protostars, and the planets derived from them, will contain grains containing light-atom molecular material with a high D/H ratio. When the collision interface reaches the deuterium-rich region around the central core of Enyo the temperature, about 3×10^6 K, is high enough to trigger a nuclear explosion, first fuelled by D–D reactions but later, as the temperature increases to 10^8 K and beyond, by other reactions involving heavier elements. Fragments of the two cores survive and are identified as the Earth, coming from the Bellona core, and Venus, coming from the Enyo core. The resisting medium is still present and the fragments round-off and decay by the Type I migration process, ending up in the terrestrial region.

Because of their masses the colliding planets should have had extensive satellite families with some members even bigger than Ganymede, the largest of the Galilean satellites. One large satellite, retained by the Earth fragment, is the Moon. It would have been in synchronous orbit around its original major planet and hence the face pointing towards the planet would also have pointed towards the collision. Debris, moving at $100 \, \mathrm{km \ s^{-1}}$ would have abraded this face and removed about a 50 km thickness of the solid surface material. Later, when the whole surface of the Moon was bombarded with large projectiles, only on the thinned side would cracks in the floors of the massive craters they produced, have penetrated deep enough to bring molten material to the surface, giving maria as seen today. This explains the hemispherical asymmetry of the Moon. The distribution of matter in the Moon, moulded by its formation around Bellona, would also ensure that it was the bombarded hemisphere that now faces the Earth.

Two large satellites are taken as having produced the smaller terrestrial planets, Mars and Mercury. Mars also shows hemispherical asymmetry, with northern lava plains and southern highlands. The angle between the spin axis and the plane of asymmetry, 56°, is due to the action of polar wander to satisfy the condition that the spin axis should lie along the principal axis of maximum moment of inertia. The high density of Mercury, higher than that of any other planet, is due to a large amount of mantle material having been removed by abrasion. The Caloris Basin and the antipodal Chaotic Terrain may be the result of the subsequent readjustment of surface material to give an equilibrium configuration.

The debris from the collision would have stretched out to a great distance, much of it, especially icy material, being lost from the Solar System. Inner core material, silicates and iron, would be closer in and after interactions with planets, the orbits of which were still evolving, and collisions with each other, a tiny amount of this would reside in the Asteroid Belt or in safe regions between major planets; an example is the large asteroid Chiron, with comet-like characteristics, between Saturn and Uranus. Some might have become attached to major planets in the form of Trojan asteroids, the rest being either expelled from the Solar System or absorbed by one of the major planets.

Icy material went further out and the effect of the resisting medium on this low-density material is to increase its perihelion. That which concluded its orbital evolution with a perihelion beyond the orbit of Neptune populated what is now known as the Kuiper Belt that, due to perturbation by Neptune, provides short-period comets. Some icy-silicate material has aphelia of tens of thousands of au and constitutes the population of the Oort Cloud that, due to perturbation by passing stars and giant molecular clouds, is the source of new comets. However, this model predicts that there is a continuous population of cometary bodies from the inner reaches of the Kuiper Belt out to the Oort Cloud.

Within the Kuiper Belt there are a number of larger bodies that, together with Ceres, are referred to as dwarf planets. These are identified as satellites of the colliding planets, the orbits of which evolved to keep their perihelia outside the orbit of Neptune. The exception is Pluto, the orbit of which at perihelion is just within Neptune's orbit. Because of the 3:2 commensurability of its orbital period with that of Neptune it never approaches that planet closely.

Neptune has a large satellite, Triton, which is in a retrograde orbit and it has been postulated that there is some event that explains the relationships, both different, of Pluto and Triton to Neptune. The interpretation, which is an outcome of the planetary collision, is that Triton was an escaped satellite with an orbit that stretched out beyond the orbit of Neptune and that Pluto had been a regular satellite of Neptune. On its journey inwards towards the Sun, Triton struck Pluto a glancing blow that propelled Pluto into its present heliocentric orbit, gave it a retrograde spin and sheared off the material to provide its comparatively large satellite Charon, with fragments providing the smaller satellites of Pluto. In the collision Triton lost energy and was captured into a retrograde orbit around Neptune that later decayed and rounded-off. A simulation supports the plausibility of this event.

A final outcome from the planetary collision is the explanation of a number of light-atom isotopic anomalies in meteorites, involving the elements carbon, nitrogen, oxygen, neon, sodium, magnesium, and aluminium.

An analysis of a nuclear explosion in material consisting of a mixture of silicates and ice, such as would have occurred in the colliding planets, explains all the observed isotopic anomalies in the above-named elements.

One last point, relates to the LHB. On the basis of the Capture Theory the Solar System would have been well settled after 500 million years. The most likely cause of the LHB is the passage of the Solar System through a giant molecular cloud that would have stripped off the major part of the comet population, especially that far from the Sun. Subsequent major disturbances of the comet population, as evidenced by a series of later bombardments were, perforce, on a much smaller scale.

This flow diagram (Figure 12.1) shows the development of the Capture Theory and the causal relationships that link various features of exoplanets and of the Solar System.

Solomon: I would like to spend a few minutes looking at your flow diagram.

They spend about 10 minutes quietly drinking while Solomon looks at Figure 12.1.

Solomon: I suppose that I ought to sum up what I think about what you have been telling me for the last three months. I have done a little research on the Web after each of your presentations in the past few weeks and what I find is that large numbers of very eminent and intelligent scientists are working on various aspects of the Nebula Theory. Although Steve has been adamant that the Nebula Theory lacks credibility, each paper I read from workers in that camp seems very respectable, scientific and convincing. The theory offers many challenges but they are all being tackled.

Conversely, the Capture Theory is the product of comparatively few people, mostly working in one location, so there is the possibility that thinking has been somewhat restricted and channelled in one direction and that other

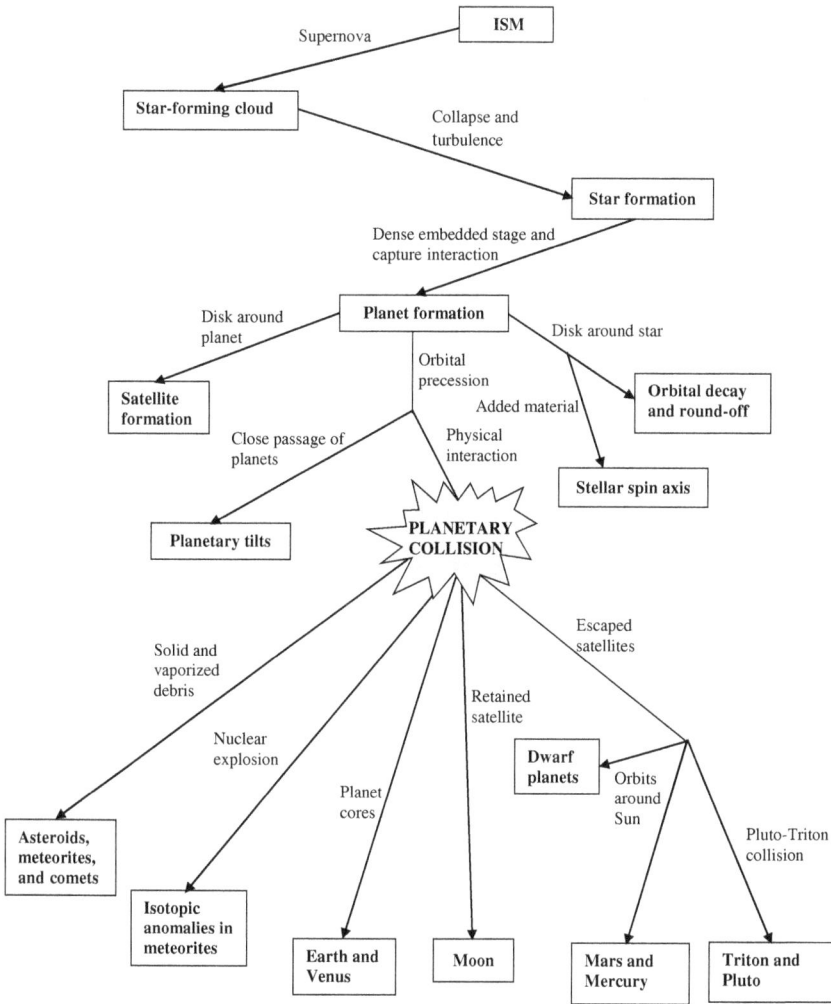

Figure 12.1   Consequences of the Capture Theory and a planetary collision

possibilities have not been considered. However, I must say
that I find the capture-theory narrative, as summarized in
the flow diagram, rather compelling; it has the continuity
of a full-length novel whereas the nebula-theory narrative
seems more like a set of loosely connected short stories,
each good in its way but lacking cohesion.

I am not sure how your subject area will develop but it seems in a very interesting state at present.

Steven: Well summarized, Sol, and very diplomatic. We're both still your friends.

Simon: Hear, hear.

Solomon: Although we've reached the end of the road as far as planetary formation is concerned it would be a pity to stop our weekly meetings. I have rather looked forward to them as a very pleasant and convivial way to finish the working week. Could I suggest that we meet next week and I'll tell you about what I am doing and how it relates to other work in particle physics.

Steven: Fine, Sol, I look forward to that but keep it simple.

Simon: The simpler the better.

They finish their drinks and depart.

Index